SLAB
SERIF
TYPE

Thames & Hudson

A·CENTURY·OF BOLD LETTERFORMS

SLAB SERIF TYPE

STEVEN HELLER AND LOUISE FILI

Steven Heller is the co-chair of the MFA Design / Designer as Author + Entrepreneur program at the School of Visual Arts, New York, and the "Visuals" columnist for *The New York Times Book Review*. He is the author, editor, and co-author of over 170 books, including fifteen with Louise Fili.

Louise Fili is the principal of Louise Fili Ltd, a New York City-based design firm specializing in restaurant identities and food packaging. The author of *Italianissimo*, she was inducted into the Art Directors Club Hall of Fame in 2004.

First published in the United Kingdom in 2016 by Thames & Hudson Ltd, 181A High Holborn, London, WC1V 7QX

Slab Serif Type © 2016 Steven Heller and Louise Fili

Designed by Louise Fili Ltd
Cover design by Louise Fili Ltd

British Library Cataloguing-in-Publication Data
A catalogue record for this book is available from the British Library

ISBN 978-0-500-51849-6

Manufactured in China by Imago

To find out about all our publications, please visit **www.thamesandhudson.com**. There you can subscribe to our e-newsletter, browse or download our current catalogue, and buy any titles that are in print.

CONTENTS

INTRODUCTION

THEY COME FAT AND THIN, ROMAN AND ITALIC, BOLD AND MEDIUM. They represent modernity, from the early nineteenth century through to the Machine Age. They were cut in wood, founded in metal and are currently digitized. They were manufactured in the United States and throughout Europe for use as both workhorse and decorative typefaces for posters, proclamations, bills, newspapers, packages, and magazines. They have names like Karnak, Memphis, Pharaoh, and Cairo, and are known by the overall umbrella designation "Egyptian." They are slab serif or square serif typefaces, and they are among the most ubiquitous and demonstrative faces in the Western world.

Slab serif faces (at least their names) derive in part from Napoleon Bonaparte's 1798 military campaign through Egypt and Syria to protect French interests and undermine Britain's access to India. The incursion, which ended in 1801, gave rise to the *Description de l'Égypte*, a series of works describing the country and its natural history. Their publication is credited with inspiring France's sudden popular wave of "Egyptomania." This fanatical and romantic obsession with ancient Egypt spread even further in 1822, when Jean-François Champollion deciphered the Rosetta Stone. Shortly afterward, these graphically intense typefaces with block serifs—perhaps a reference to the massive stones that form the pyramids—were marketed in nineteenth-century France under the family name Egyptiennes and helped define the graphic styles of the era.

Slab serifs also represent the Machine Age. According to the Maximilien Vox Association Typographique Internationale (ATypI) classification system, slabs are also called mechanical or *mécanes*, a name that refers to their genesis during the Industrial Revolution and later shifts in production during the early twentieth century. A 1930s version is aptly called "Girder," a reference to the skyscraper construction wave. The principal characteristics of these typefaces are a very low contrast and rectangular slab serifs, corresponding to the Egyptiennes of the Thibaudeau classification, which groups typefaces into four families according to shape and serif appearance. This category includes both typefaces with bracketed serifs (Clarendons) and typefaces with square or unbracketed serifs (Memphis).

The above classifications were established in modern times, during the early 1920s, yet slabs have a much longer legacy. The earliest recorded slab serif typeface for job printers, Antiqua, was introduced in 1815 by the legendary Vincent Figgins foundry in London. The term Egyptian had previously been used to describe sans serif types in England. But whatever they were called, slab serif typefaces filled the streets in the birthplace of the Industrial Revolution, dominating the new medium of advertising posters. In return for their investment, advertisers demanded an extra visual impact that these attention-grabbing slabs effectively provided. Predictably, slabs became common currency for demonstrative communications.

The slabs spawned a rash of variations and sub-classifications that are all represented in some way in this book:

CLARENDONS (the original designed by Robert Besley for Thorowgood and Co.) are distinguished from other slab serifs by their bracketing and some contrast in size in the actual serif.

NEO-GROTESQUES are the most common slab serif typefaces, with evenly weighted stems and serifs.

ITALIENNE typefaces, where the serifs are even heavier than the stems, are a familiar sight in the United States on Western "Wanted" posters.

TYPEWRITER slab serifs originated in monospaced format with a fixed width, where every character takes up the same amount of horizontal space.

Some slabs are anonymous, orphaned without a name, but others are renowned for their utility and versatility. One such, Memphis, was designed at the D. Stempel Foundry in 1929 by Dr. Rudolph Wolf. His was the first revival during his time of the Egyptian slab serifs. Memphis quickly became one of the most popular typefaces of the period and began appearing in commercial print around the globe. As with the earlier Egyptian fonts, its high legibility and even weight values make the font highly effective in brand and display use. Its creation led to an extensive revival of slab serifs in foundries around the world.

There are dozens more faces and fonts that have been published in varying degrees of frequency. This book pays homage to their functionality and beauty, quirkiness and versatility, through type specimens and hundreds of stunning applications. It is a paean to all who designed them and inspired them. Merci, Monsieur Bonaparte.

AMERICAN

TYPOGRAPHIC PUBLIC NOTICES PRINTED IN THE UNITED STATES during the late eighteenth century were styled after book title-page layouts, with "polite" Latin typefaces in central axis compositions. In the early nineteenth century a sea change occurred as a new generation of type emerged, called slab or square serifs. According to Elizabeth M. Harris, who wrote about nineteenth-century wood types for the Smithsonian Institution's National Museum of American History, this fashion began in England with the genre known as fat faces. These typefaces had stouter proportions than more nuanced old style "romantic" letters. "'Fat face,' in the strictest sense, refers to letters developed from the eighteenth-century types of Didot and Bodoni," Harris wrote. "To eyes accustomed only to the old style, the new letters were grotesque and offensive." Nonetheless, fat faces came to define the national type style in America.

While most early display letters were cast in lead, in the 1820s hand-cut wood type offered larger character sizes than metal could possibly achieve. Initially, there were no mass-production tools, which made wood type costly. American inventors, most notably Darius Wells, who launched a wood type company in 1828, quickly came to the rescue with machines such as his lateral router, which cut away the background wood to leave a sharp printable image. Other sophisticated machines followed. Heavy slab typefaces came about, in part, because they were well suited to this technology. Another important wood type-cutting device, the pantograph, followed a master pattern that enabled simultaneous cuttings with more size options, making it possible to expand and condense letters. The quintessential American slabs have exaggerated serifs, but others were also available in America, including English and French "Antique" and "Egyptian" letters. Slab wood type was mainly used for posters, but when made in metal and film, slabs were used for almost every printing need, old-fashioned and new. Dan Smith wrote in *Graphic Arts ABC* that the modern square serif design performed "a highly useful function in tying letters together into words and making the word units easier to recognize." Moreover, he added, these types have "made an epochal contribution to the *matériel* in our typographic armory."

AMERICAN JUBILEE

NEW YORK WORLDS FAIR 1940

15¢

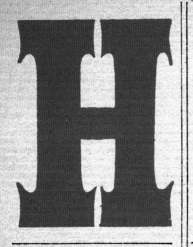  

PRICE PER LETTER.		PRICE PER LETTER.		PRICE PER LETTER.		PRICE PER LETTER.	
Holly Wood.	Wood Type.	Holly Wood.	Wood Type.	Holly Wood.	Wood Type.	Holly Wood.	Wood Type.
$0.03	$0.06	$0.03	$0.05	....	$0.06	$0.03	$0.05
.03	.06	.03	.05	$0.03	.06	.03	.05
.03	.07	.03	.05	.03	.07	.03	.05
.04	.09	.04	.07	.04	09	.04	.07
.04	.11	.04	.09	.04	.11	.04	.09
.05	.13	.05	.10	.05	.13	.05	.10
.05	.15	.05	.11	.05	.15	.05	.11
.05	.16	.05	.12	.06	.16	.05	.12
.06	.17	.06	.13	.06	.17	.06	.13
.06	.18	.06	.14	.07	.18	.06	.14
.07	.19	.07	.15	.08	.19	.07	.15
.08	.22	.08	.17	.10	.22	.08	.17
.09	.23	.09	.18	.10	.23	.09	.18
.10	.26	.10	.21	.12	.26	.10	.2
.12	.28	.12	.23	.15	.28	.12	.23
.15	.30	.15	.25	.17	.30	.15	.25
.17	.32	.17	.27	.18	.32	.17	.27
.18	.35	.18	.30	.18	.35	.18	.30
.21	.41	.21	.35	.21	.41	.21	.35
.25	.48	.25	.42	.25	.48	.25	.42
.30	.60	.30	.55	....	.60	.30	.55

LETTER.

Price Per Letter.		Price Per Letter.		Price Per Letter.	
Holly Wood.	Wood Type.	Holly Wood.	Wood Type.	Holly Wood.	Wood Type.
$0.02	$0.06	$0.02	$0.05	$0.02	$0.05
.03	.07	.03	.05	.03	.05
.04	.08	.03	.06	.03	.06
.05	.10	.04	.08	.04	.08
.96	.12	.05	.10	.05	.10
.07	.14	.05	.12	.05	.12
.07	.16	.06	.14	.06	.14
.08	.18	.07	.15	.07	.15
.08	.20	.07	.16	.07	.16
.09	.22	.08	.17	.08	.17
.10	.24	.09	.18	.09	.18
.13	.28	.12	.21	.12	.21
.13	.30	.12	.22	.12	.22
.15	.38	.14	.25	.14	.25
.18	....	.16	.28	.16	.28
.20	....	.18	.30	.18	.30
.23	....	.20	.33	.20	.33
.25	....	.20	.36	.20	.36
....	....	.25	.44	.25	.44
....	....	.30	.52	.30	.52
....	....	....	.65	....	.65

The GAY NINETIES

Nostalgic designs redolent of a tenderly-remembered period in America . . and still increasing in current popular favor.

BEEF TRUST, No. 1 and No. 2

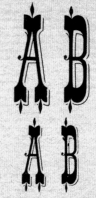

ELEGANTE, No. 1 and No. 2

HANDSOME
Nos. 1 - 2 - 3

SHOWBOAT, No. 1 and No. 2

HORSECAR, No. 1, No. 2, No. 3

BACK BAY, No. 1 and No. 2

MUSIC HALL, No. 1 and No. 2

RUSSELL, No. 1 and No. 2

PHINEAS, No. 1 and No. 2

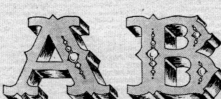

ROCOCO

The GAY NINETIES

SUMPTUOUS, No. 1 and No. 2

VALENTINE, No. 1 and No. 2

CROQUET
No. 1 and No. 2

Daring, dynamic, designs that impart a definite, dazzling, diversity to your layout. Exclusive with Franklin Typographers.

REGAL

GIBSON GRAY

TEMPERANCE, No. 1 and No. 2

GIBSON OPEN

GIBSON SHADED

ROSEBUD, No. 1 and No. 2

TINTYPE, No. 1 and No. 2

GIBSON BLACK

EMPORIUM

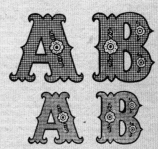

HARPER'S, No. 1 and No. 2

YEAGER MILLING CO'S

STRAIGHT

PATENT

TRADE-MARK.

ALFARATA

YEAGERTOWN, PA.

THE COMIC
ALMANAC,

OLD Mr. 1874.   Permit me to Introduce your Successor,   YOUNG Mr. 1875.

FOR THE YEAR 1875.

BEHM & GERHART,
BOOKSELLERS AND STATIONERS,
No. 305 Market Street,
PHILADELPHIA.

DEMO...

...USTRATED

SATURDA...

HARPER'S

JOURNAL O...

RESTS MONTHLY

GLOBE.

JULY 1, 1899.

WEEKLY

CIVILIZATION

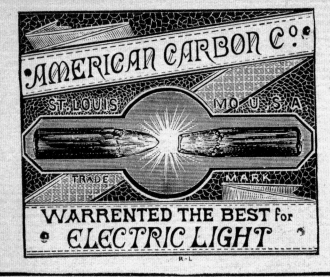

18

R. TUTTLE & CO.,

Fruit and Produce Commission Dealers,

43. NO. MARKET ST., BOSTON.

From G. M. NICHOLS, Weir, Mass.

SUMNER N. STAPLES

MYRICKSVILLE, MASS.

Breeder and Shipper of
HIGH-CLASS Thorough-Bred
Wyandottes, P. Rock,
W. F. B. Spanish,
W.C.B.Polish, L. Brahmas,

After Eggs for hatching.
PEKIN DUCKS.
...se and Single Comb, White and Brown Leghorns.
EGGS $1.00 per 13.

S. CHAMBERLINE & CO.
BOSTON.

FROM A. J. GIDMARK,

FGSR

MGDN

DKXUI

PSM

NHJGM

LFRH

THADDEUS DAVIDS & CO.

127 & 129 WILLIAM ST.

STEEL PEN •

INK

BLACK

FOR

WRITING

RECORDS

MANUFACTORY ESTABLISHED 1825.

EXCELSIOR.

the Fairs of the
... and the
... Institutes
... Silver Medals
... Diplomas
...warded to
...tor for the
... Inks,
... Wafers &c.

This Ink possesses
all the qualities
required for a good
Writing Ink
and however pale
may appear when
first opened it will
soon change to
Deep Black

...ANTED

NEW YO...

...the Act of Congress in the year 1853 by Thaddeus Davids in the Clerks Office of the District Court of the Southern ...

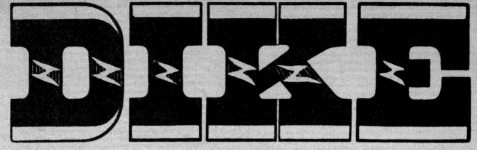

ANTISTROPHE

SEISMOLOGY

Lizar DBir DSeal

IGNEOUS GLACIER HOT

15 Halibut Cove 75

1$ TRAJECTORY

ALBERTA

Burpee's Seeds Grow

BURPEE'S
NETTED GEM
OR
ROCKY FORD

THE
MELON THAT MADE
ROCKY FORD
FAMOUS

W. Atlee Burpee
Seed Growers, Philade

Burpee's Annual

BURPEE'S
BLUE BANTAM
PEA

THE PLAIN
TRUTH ABOUT
SEEDS
THAT GROW

1919

BURPEE'S
CHINESE GIANT
PEPPER

W. Atlee Burpee Co.,

Seed Growers   Philadelphia.

ENDIVE

BROAD LEAVED
ESCAROLE SCHIANI

KOHL RABI

EARLY WHITE VIENNA

WISS CH

SEA KALE OR F

KALE

DWARF CURLED SCOTCH

RADISH

EARLY WHITE TURNIP

CARD SE

FREDONIA, N

CABBAGE

AMERICAN SAVOY
VERZE

CARD SEED.CO.
FREDONIA, N.Y.

EED.CO.
N.Y.

GENESEE VALLEY LITHO. CO., ROCHESTER, N.Y.

29

EL RIO ORCHARD

BRAND

CONTENTS
4/5 BUSHEL

California *Bartlett*

PEARS

PRODUCE OF U.S.A.

DiGiorgio
FRUIT
CORPORATION
SAN FRANCISCO CALIF.

GOLETA

BRAND

famous
Santa Barbara
County

LEMONS

Sunkist

GOLETA LEMON ASSOCIATION
GOLETA ··· CALIFORNIA

GROWN IN U.S.A.

30

S·L
BRAND

GROWN AND   PACKED BY
THE SUNLAND PACKING HOUSE CO.
PORTERVILLE · TULARE COUNTY · CALIFORNIA

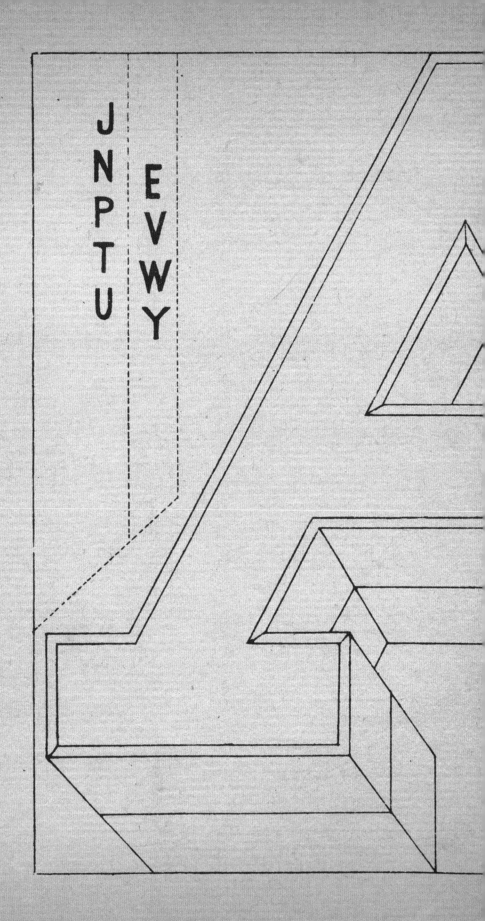

Palm's Patent Transfer Letters.

J N P T U

E V W Y

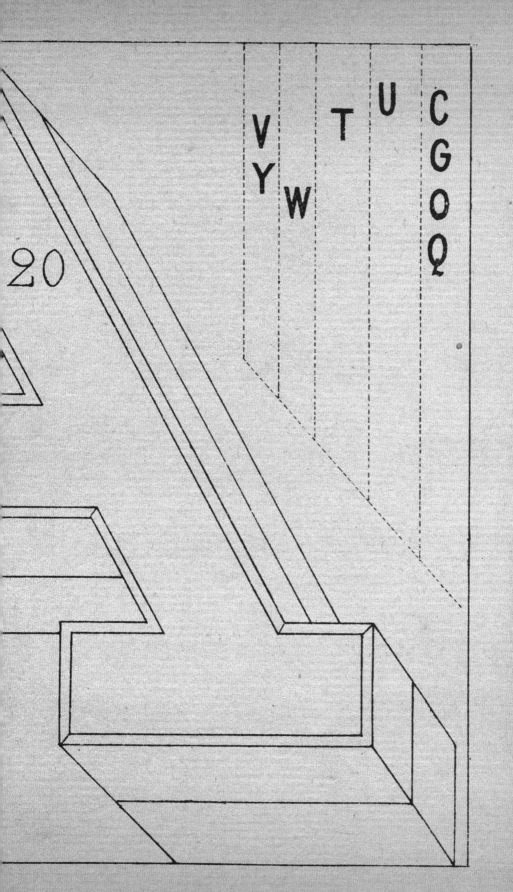

No. 20   Executed in Colors in following styles: **D. E. J. K. O & P.**   **S & T.** see Color Plate I & II.

No. 20.   For Glass, executed in Color styles: **CC & HH.**   see Color Plate III.

SNELLEN'S TEST TYPES

FEET		METRES
200	M	61.00
100	T E	30.50
70	L E H	21.34

W-75, 10 Line — Caps, figures

FURNITURE POLISH 3456

W-76, 8½ Line — Caps, lower case

TAB GUIDES

W-77, 8 Line — Caps

LOBSTER & CLAM

W-78, 8 Line — Caps, figures

wratte 983 filter

W-79 8 Line — Caps, figures

SOFT SHELL CRABS

W-545—10 Line C, lc, F, P

TOP is BACK

W-546—7 Line C

BAGEL & LOX

W-547—8 Line C, lc, P

MEAT is GOOD

W-548—12 Line C, F

CAPE

W-549—10 Line C, F, P

VIKINGS AR

37

The
AMERICAN PRINTER

Craftsmen's Number

NINE EAST THIRTY-EIGHTH STREET · NEW YORK CITY

Volume 93 ⩝ Number 2 ⩝ 25 Cents a Copy ⩝ $3.00 a Year

AUGUST · 1931

Letterheads

A DEMONSTRATION OF THE NEW EAGLE·A CONTRACT BOND

THEN AND NOW

INQUEST
Before the Honorable Lee Slocum,
Coroner, and the Coroner's Jury, on this
"ONE OF THE SIX BEST DETECTIVE STORIES OF THE YEAR" — LONDON TIMES.
PERCIVAL WILDE

THE TEXTBOOK

OF THE FUTURE AND ITS FORERUNNERS

AN EXHIBITION

FEBRUARY 6-28, 1936

THE NATIONAL ARTS CLUB

15 GRAMERCY PARK, N. Y.

THE AMERICAN INSTITUTE OF GRAPHIC ARTS

PULLM

makes square slices

easier to handle!

STEINECK BA
UNIONTOWN,
16 OZ. OR M

PULLN

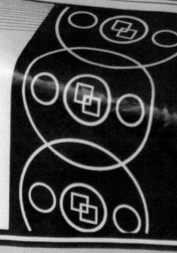

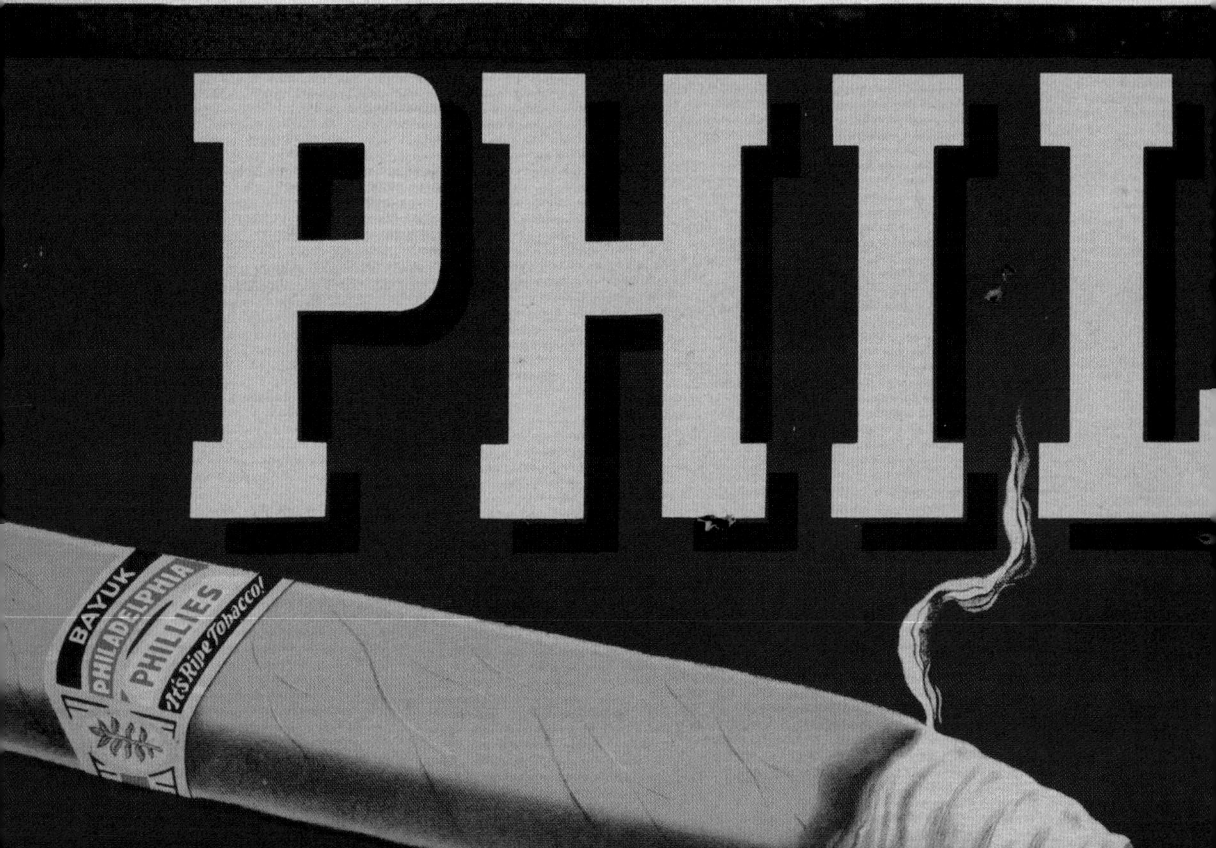
PHIL
BAYUK
PHILADELPHIA
PHILLIES
It's Ripe Tobacco!

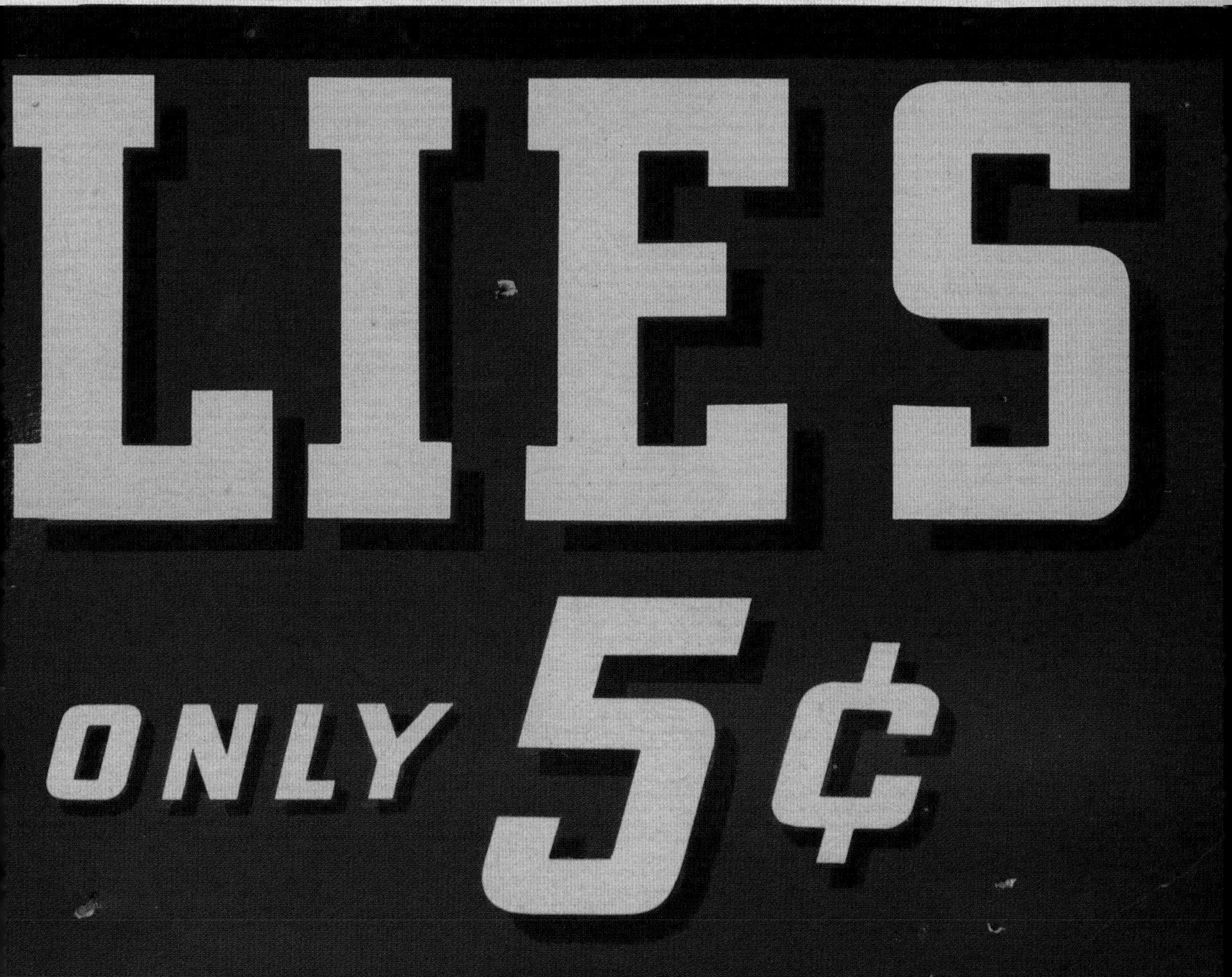
LIES
ONLY 5¢

DIE FLEDERMAUS

STUDENT PRINCE

OPERETTA

marek weber a

GYPSY BARON

COUNTESS MARITZA

POTPOURRI

d his orchestra

Europe
in
Revolt

RENÉ KRAUS

MORE BUSINESS

Smythe 49

-LEO RACKOW

HOPALONG CASSIDY RADIO

OPERATING INSTRUCTIONS

OKAY PARDNER YOU'RE ALL READY FOR A ROUND UP!

Whistle up the boys and let 'em have a look. Pardner, you're a top hand, now. All fixed up. You've got a real, range ridin' radio—best in the business.

Hoppy

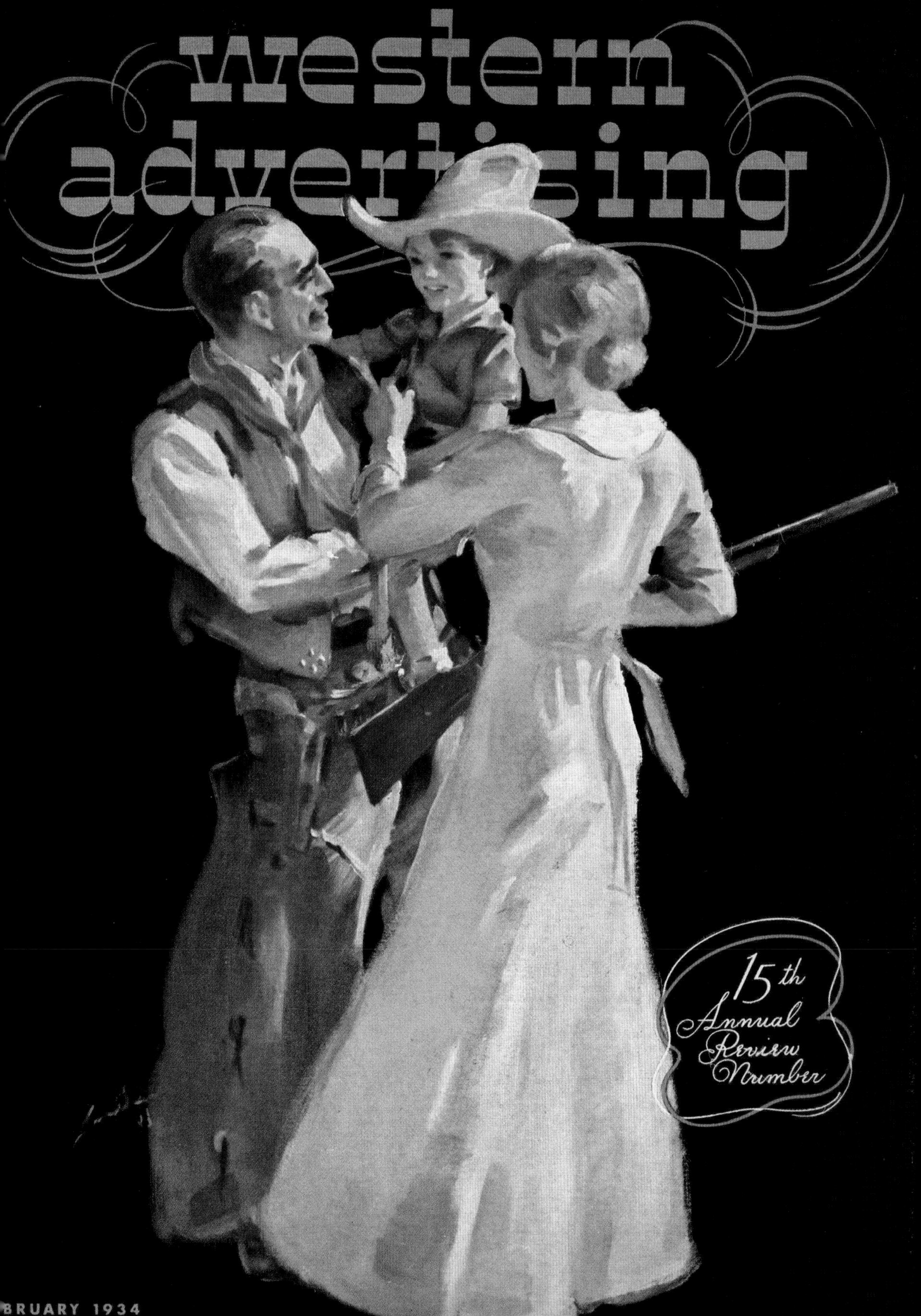

BRUARY 1934

120 Point ★   3A $44 25

BIG

42 Point   5A $9 25

MONEY

36 Point   6A $8 00

FASHION

96 Point ★   3A $28 50

SUN

30 Point   7A $6 00

HUMIDORS

24 Point   8A $5 00

SUPERIORITY

84 Point ★   3A $22 00

HIDE

72 Point ★   3A $17 75

BARK

Characters in Complete Font

A A A B C
D E F G H I
J K L M N
O P Q R R
S T U V W
X Y Z & $
$ 1 2 3 4 5
6 7 8 9 0 0
. , ' ' : ; ! ?

The following characters are furnished mortised with all sizes

A F L P T
V W Y

60 Point   4A $14 50

MINDS

54 Point   4A $12 25

RICHLY

Not sold in weight fonts

★ Sizes 72 to 120 Point, any assortment of words or letters, will be furnished by the inch. Minimum, 8 inches of any one face and body

THE AMERICAN PRINTER

VOLUME NUMBER NINETY-EIGHT | ISSUE NUMBER FIVE

for MAY

Price 25 Cents a Copy
$3.00 a Year

ADVERTISING·ARTS

SEPTEMBER·1934
PRICE 50 CENTS

FOCUS

10¢
APRIL

BIRTH
OF A
DUMMY

THE NEW ART

OF

Modern Cooking

EVERYTHING for the GARDEN

1937

THE JOHN HANCOCK GARDEN — SEE PAGE 2 OF COVER

PETER HENDERSON & CO.
35 CORTLANDT STREET
NEW YORK

ABCDE
GHIJK
LMNOP
QRSTU
VWXYZ
ABCDEFG

61

ABC[
HIJ[
NOP[
UVW

DEFG
KLM
RST
XYZ

Town & Country

ESTABLISHED 1846

66

Raoul Dufy

JANUARY, 1937
PRICE 50 CE

Town & Country

ESTABLISHED 1846

NOVEMBER, 1935
PRICE 50 CENTS

67

JANUARY, 1943

1

architecture   civic design   the arts

Typical of the products of the future is magnesium, which has a tremendously high strength-to-weight ratio. Not commercially available in the past, it is now being produced in quantity, from sea water, for armaments. What will happen to building design when our newly achieved capacity to produce magnesium is channeled into peacetime uses? This is a photomicrograph of part of a magnesium casting for an aviation engine.

NEW PENCIL POINTS

THEATRE

complete play "LIGHT UP THE SKY" by Moss Hart

NEW PENCIL POINTS

NEW PENCIL POINTS

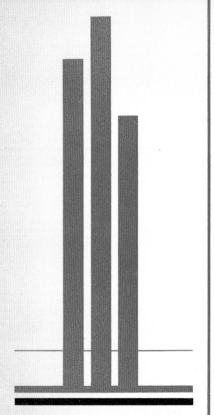

A B C D E F G
H I J K L M N
O P Q R S T U
V W X Y Z & $
1 2 3 4 5 6 7 8 9 0
a b c d e f g h i
j k l m n o p q r
s t u v w x y z
. , - ' ' : ; ! ?

The following characters are cast recessed
from 30 point to 72 point

A F L P T V W Y

Tower Series

72 Point   3 A 5 a

UNIQUE Styles

60 Point   5 A 8 a

Latest SPECIMEN

48 Point   6 A 10 a

DISTINCTIVE MODES

Advertising Campaign

36 Point   7 A 11 a

SMARTLY PRINTED BOOKS

Brilliant Biography Displayed

30 Point   8 A 13 a

FOREIGN AGRICULTURE REVIVED

English Merchants Unusually Joyful

24 Point   10 A 20 a

INCREASED SALES

Delights Executives

14 Point   19 A 38 a

MODERNISTIC FURNISHINGS
Designer Helps Brilliant Artist
Received Thorough Instruction

18 Point   15 A 30 a

MARVELOUS PICTURES

Blending Shape Harmony

12 Point   25 A 49 a

SMASHING STROKES PEP DISPLAY
Brilliant Italian Lithographer Arrives
Exquisite Specimens Were Displayed

Photo

engraving

engraving

engraving

engraving

engraving

CAIRO

● AN INTERTYPE FACE

NORRIS DAM

TENNESSEE VALLEY AUTHORITY

GIRDER LIGHT

GIRDER HEAVY

TYPEFOUNDERS ASSOCIATION INC · 216 EAST 45TH STREET · NEW YORK CITY

GIRDER

WITZER

YMBOLIZES PROGRESS...

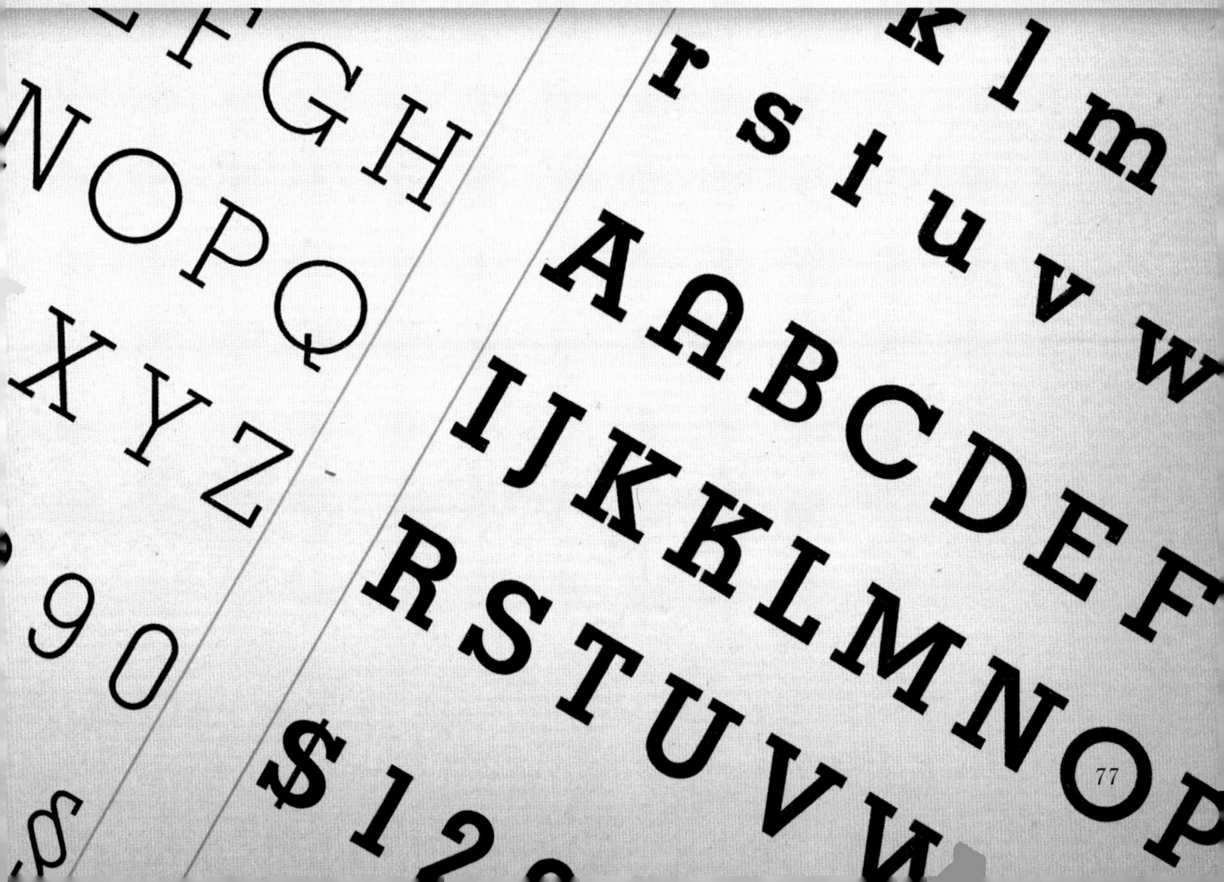

E F G H
N O P Q
X Y Z
9 0
$ 1 2

k l m
r s t u v w

A A B C D E F
I J K K L
R S T U V W M N O P

42 Point · 3A $5 90 · 5a $4 25 · $10 15

CHOICE DESIGNS
Explorers bringing
some helpful data

30 Point · 4A $3 40 · 9a $3 60 · $7 00

BUILDINGS INSURED
Local brokers offering
policies at lower rate

18 Point · 9A $2 50 · 16a $2 90 · $5 40

SPLENDID BANK REPORT
Optimistic statement shows
that business is improving
in all parts of the country

12 Point · 15A $2 10 · 28a $2 50 · $4 60

ORDER DISTINCTIVE TYPES

Any printer handling jobs that
require strong display knows
the value of a type that was
designed with this in mind

8 Point · 19A $1 70 · 37a $2 10 · $3 80

RICHER CROPS EXPECTED

The reports from growers of
wheat seem to indicate that
unusually attractive prices
will prevail in all markets
during the ensuing year

A
D
H I
N O
U V W
1 2 3 4
a b c d
l m n o
w x y z
Charact-

48 Point

ARMY
Greatest

60 Point

BRING
Pleasing

AMERICAN TYPE F

Rockwell Antique may be obtained through any of our Twenty-five Selling Houses C

ANTIQUE

Cast Standard Line, Point Body and Point Set

36 Point   3A $4 05   7a $4 05 $8 10

INCREASES DEMAND
Good type impels buyer to order more printing

24 Point   5A $2 70   11a $3 10 $5 80

BUSINESS MANAGEMENT
Merchants who are striving to succeed should learn the importance of advertising

14 Point   12A $2 30   22a $2 70 $5 00

MODERN TYPOGRAPHIC STYLE
Attention of all printers is called to this excellent type face especially adapted to the vogue which now prevails in display composition

10 Point   16A $1 90   30a $2 30 $4 20

EXCELLENT RESULTS SECURED
Numerous workmen throughout the nation are now studying personal efficiency, while their employers are encouraging the movement by giving it generous support

6 Point   21A $1 60   40a $1 80 $3 40

MAKE NEW YEAR RESOLUTION
During the year every compositor should create a piece of printing that embodies his knowledge of good typography for no profit except the noble one of great pleasure taken in this work

---

G
L M
R S T
Z & $
7 8 9 0
h i j k
s t u v
' : ; ! ?
...ete Font

3A $7 25   4a $4 35 $11 60

TEAM
Strength

3A $11 35   4a $6 55 $17 90

MINER
Designer

UNDERS COMPANY

...tly Located in Principal Cities . . . . Sold in Single Fonts and also in Weight Fonts

...SED   Amtyco No. 25

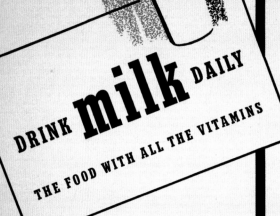

DRINK **milk** DAILY

THE FOOD WITH ALL THE VITAMINS

price list

Lida CANDID CAMERA CORP.

PLEASE SEND SAMPLES OF **"DOUBLE-O"** BOND PAPER

NAME

ADDRESS

CITY

STATE

BUSINESS

Here this versatile new face helps a cut-out, a booklet cover and a mailing card get over their very different messages quickly and effectively

STYMIE BOLD CONDENSED

ABCDEFGHIJKLMNOPQRSTUVWXYZ&abcde
fghijklmnopqrstuvwxyz$1234567890.,-`':;!?

72 Point 3 A $6.40 4 a $3.10 $9.50

DRAMAT
Dramatiz

60 Point 3 A $4.35 6 a $4.55 $8.90

DRAMATIZ
Dramatizet

48 Point 5 A $4.30 9 a $4.70 $9.00

DRAMATIZE Y
Dramatize you

36 Point 6 A $3.10 11 a $3.40 $6.50

DRAMATIZE YOUR
Dramatize your me

30 Point 7 A $2.50 14 a $3.10 $5.60

DRAMATIZE YOUR M
Dramatize your mess

24 Point 9 A $2.10 17 a $2.55 $4.65

DRAMATIZE YOUR MESS
Dramatize your message

18 Point 13 A $2.00 25 a $2.35 $4.35

DRAMATIZE YOUR MESSAGE B

Dramatize your message by setting it effectively in an approp

14 Point 17 A $1.90 33 a $2.10 $4.00

DRAMATIZE YOUR MESSAGE BY SETTI

Dramatize your message by setting it effectively in an appropriate, legible t

12 Point 20 A $1.70 40 a $2.00 $3.70

DRAMATIZE YOUR MESSAGE BY SETTING IT

Dramatize your message by setting it effectively in an appropriate, legible type face.

81

72 point:

NEW STUDENT 12
Earns high honors

60 point:

HISTORICAL GROUP 41
Meets in new building

48 point:

ANNUAL GOLDEN GLOVE 86
Bouts to be held in February

42 point:

RADIO STARS GIVE BENEFIT 57
Performance at the new stadium

36 point:

PRINTING EXHIBIT OPENS TODAY 90
To include many old and rare books

30 point:

JOURNALISTS' MEETING 25
Discusses new code of ethics
governing future foreign acts
to further insure world peace

24 point:

HEAVY SNOW STORM 89
Ties up all traffic for days

18 point:

MERCHANTS SPRING OUTING 65
Attracts a good many new friends

LUDLOW TYPOGRAPH COMPANY · 2032 CLYBOURN AVENUE · CHICAGO · ILLINOIS

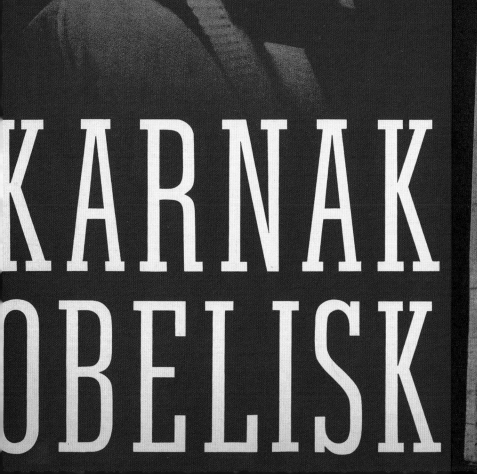

LUDLOW

KARNAK
OBELISK

BETON

MEDIUM • BOLD • EXTRA BOLD • INTERTYPE

BETON · BOLD

The architectural element has been emphasized in most cases by simple wall surfaces, subdued color schemes and an absence of all superfluous DETAILS THAT DISTRACT THE EYE FROM THE

Once the sculptor had finished the original model, this was purchased by the factory and passed on to great CAMBRIDGE EDITIONS RESTORED

Small bookcases let in the wall are a decorative feature of not GREAT EUROPEAN HOSPITAL

They are well aware that to the laymen the extra is A SIGN OF INADEQUATE

It is again instructive to recall the changing HEROIC REGIMENTS

There is a danger, too that the industry is NEWFOUNDLAND

Master of French HONEST CODE

Savoy Hotel BRIGHTON

Cover Design INSPECTION

Defendant MENDING

Uniform ORGAN

Record

Drake

BETON · EXTRABOLD

14 Point   19A 37a

Division of Graphic Arts
Printing Crafts Building
ADVERTISING AGENCY

16 Point   15A 32a

Historical Portraits
New Incorporation
PRECIOUS STONES

18 Point   12A 22a

Tip on the Market
Strathmore Paper
OUTBOARD RACE

24 Point   8A 15a

Flower Garden
Inspired Singer
VISIT EUROPE

30 Point   6A 12a

Merchants
PROBLEM

36 Point   5A 10a

Dispatch
BREACH

48 Point   5A 9a

Furniture
SERVICE

60 Point   5A 8a

Garden
NORTH

72/60 Point   4A 6a

Fabric
BANK

84/72 Point   3A 4a

Hotel

Continental Typefounders

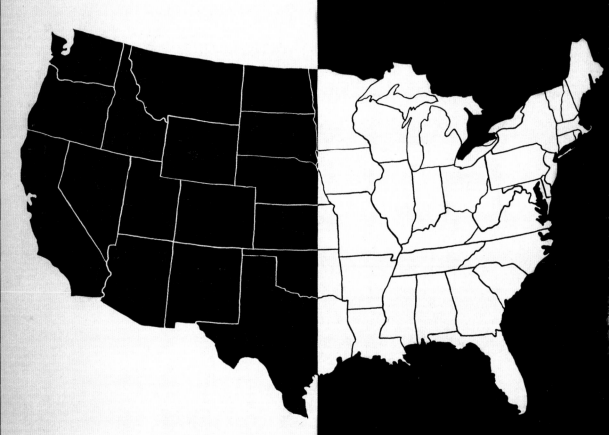

CITY

COMPACT NORMAL   COMPACT BOLD

PLAYBILL

FRENCH ANTIQUE *introduced into English printing during the heyday of* VICTORIAN FANTASY

TWENTY-FOUR POINT

The main purpose of letters is the practical one of making tho ughts visible. Ruskin says that "all letters are frightful things ABCDEFGHIJKLMNOPQRSTUVWXYZ&?!$1234567890

THIRTY-SIX POINT

The main purpose of letters is the practic al one of making thoughts visible. Ruskin ABCDEFGHIJKLMNOPQRSTUVWXYZ?!&$

FORTY-EIGHT POINT

The main purpose of letters is t he practical one of making thou ABCDEFGHIJKLMNOPQRSTUV! WXYZ&$1234567890

SEVENTY-TWO POINT

The main purpose of ABCDEFGHIJKLMNO

Many of the streets are broad with large houses

3778   10 point   25 A 45 a

It is surrounded by a ring of great towns

3779   12 point   18 A 35 a

Sheffield is the place of hardware

3780   14 point   15 A 30 a

National Woolprice Stabilization

3781   16 point   13 A 25 a

A name that assures quality

3782   18 point   10 A 20 a

London and its Bridges

3783   24 point   7 A 14 a

Sport and Pastime

3784   30 point   6 A 12 a

Newest Models

3785   36 point   5 A 10 a

Diametrical

3786   42 point   5 A 8 a

Breakfast

3787   54 point   4 A 7 a

General

3788   72 point small face   4 A 5 a

Action

3789   72 point large face   3 A 3 a

90

TO REDUCE THE SALES EXPENSE USE PRINTING

NEW IDEA TO REDUCE CUTTING COSTS

SAMPLE BOOK FOR EVERYBODY

IMPORTANT AND PERTINENT

NEW STEEL FURNITURES

DINING ROOM SUITES

MULTIPLICATION

STUDEBAKER

PASS-BOOK

FACTORY

INDIAN

HOME

figures: **1 2 3 4 5 6 7 8 9 0**

TIMELY·CLOTHES
3740 18 point 10 A

STAR GELATIN
3741 20 point 8 A

RYPE FRUITS
3742 24 point 7 A

DESIGNER
3743 30 point 6 A

FLOWER
3744 36 point 5 A

HEATH
3745 48 point 4 A

figures: 1 2 3 4 5 6 7 8 9 0

MEMPHIS LUNA

PROMENADE
4685 24 point 7 A

LETTERBOX
4686 30 point 6 A

GAZELLE
4687 36 point 4 A

SAILOR
4688 42 point 3 A

HAND
4689 54 point 3 A

figures: 1 2 3 4 5 6 7 8 9 0

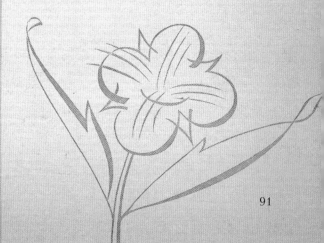

91

What you want from your car is the best performance in

3623 8 point 25 A 50 a

EVENING CLASS OF COMMERCIAL SCHOOL WEST

For those who have never taken the time to

3624 10 point 25 A 50 a

ADVERTISERS FOR BIG DISPLAY LINES

Before they had penetrated very far

3625 12 point 20 A 40 a

FACTORIES IN OKLAHOMA CITY

General Impressions of Bristol

3626 14 point 17 A 34 a

LEEDS AND BIRMINGHAM

National Lead Company

3627 16 point 14 A 28 a

ENGLISH COALFIELDS

Use and Care of Metal

3628 18 point 10 A 22 a

CATALOGUE PAGES

Modern Invention

3629 24 point 7 A 14 a

KANSAS HOTEL

Personal Matter

3630 30 point 6 A 12 a

GRAPHIC ART

Queen Mary

3631 36 point 4 A 6 a

BALTIMORE

California

3632 42 point 3 A 5 a

REVENUE

Remark

3633 54 point 3 A 4 a

MARKS

figures: 1 2 3 4 5 6 7 8 9 0

HEINZ

TOMATO KETCHUP

PURE FOOD PRODUCTS

H. J. HEINZ COMPANY, PITTSBURGH

144/120 Point   Sold by weight—3A 6a, about 95 pounds, $64 60 ★   All characters excepting descenders cast on 120 point body

Blue INK

120/96 Point   Sold by weight—3A 6a, about 60 pounds, $42 00 ★   All characters excepting descenders cast on 96 point body

FINE Ships

96/84 Point   Sold by weight—3A 6a, about 45 pounds, $32 40 ★   All characters excepting descenders cast on 84 point body

Stylish RING

84/72 Point   3A $14 30   4a $9 25 $23 55 ★   All characters excepting descenders cast on 72 point body

NICE Lingerie

72/60 Point   4A $12 05   5a $9 00 $21 05 ★   All characters excepting descenders cast on 60 point body

Displays BONDS

★ Sizes 72/60 to 144/120 Point, any assortment of words or letters, will be furnished by the inch. Minimum, 8 inches of any one face and body

AMERICAN TYPE FOUNDERS COMPANY

THE LAST PURITAN

A Memoir in the form of a Novel

GEORGE SANTAYANA

MARRIAGE

MARTHA GRAHAM

96/84 Point   Sold by weight—3A 5a, about 45 pounds, $32 40 ★

Study

84/72 Point   3A $17 00   4a $10 25 $27 25 ★

FINE

72/60 Point   4A $14 10   5a $9 15 $23 25 ★

Giganti

All characters excepting descenders cast on 84 point body

ART

All characters excepting descenders cast on 72 point body

Report

All characters excepting descenders cast on 60 point body

TASK

No. 56D (1)—Antique Pointed (No A)   36 pt.

TYPES 5
to be very

No. 56E—Antique Pointed No. 2   36 pt.

PRETTY ?
or will one!

No. 57A—Broadgauge (no fig. 2 ; caps only)   6 pt.

MOST IMPRESSIVE AND $34

No. 57B—Broadgauge (caps only)   8 pt.

ATTRACTIVE WE SAY 1

No. 57C—Broadgauge (caps only)   10 pt.

BUT NATURALLY 2

No. 57—Broadguage (caps only)   12 pt.

WE DO NOT TRY

No. 57D—Broadgauge (caps only)   18 pt.

CHIEFLY 3

No. 702—Egyptian Extended Bold (large font)   10 pt.

SET IN THE OLD-STYLE 9
Manner . . . Then, Maybe The

No. 703—Egyptian Extended Bold   12 pt.

VERY NEXT WEEK 12
Scarcely Any Orders Come

No. 704—Egyptian Extended Bold (large font)   18 pt.

IN AT ALL. WE'VE
Never Been Able To

No. 705—Egyptian Extended Bold (no l.c. g or i)   24 pt.

FIND THE 35
Because. Some

No. 706—Egyptian Extend Bold   48 pt.

TIMES,

No. 58D—Broadgauge Shaded (caps only)   36 pt.

NOW -

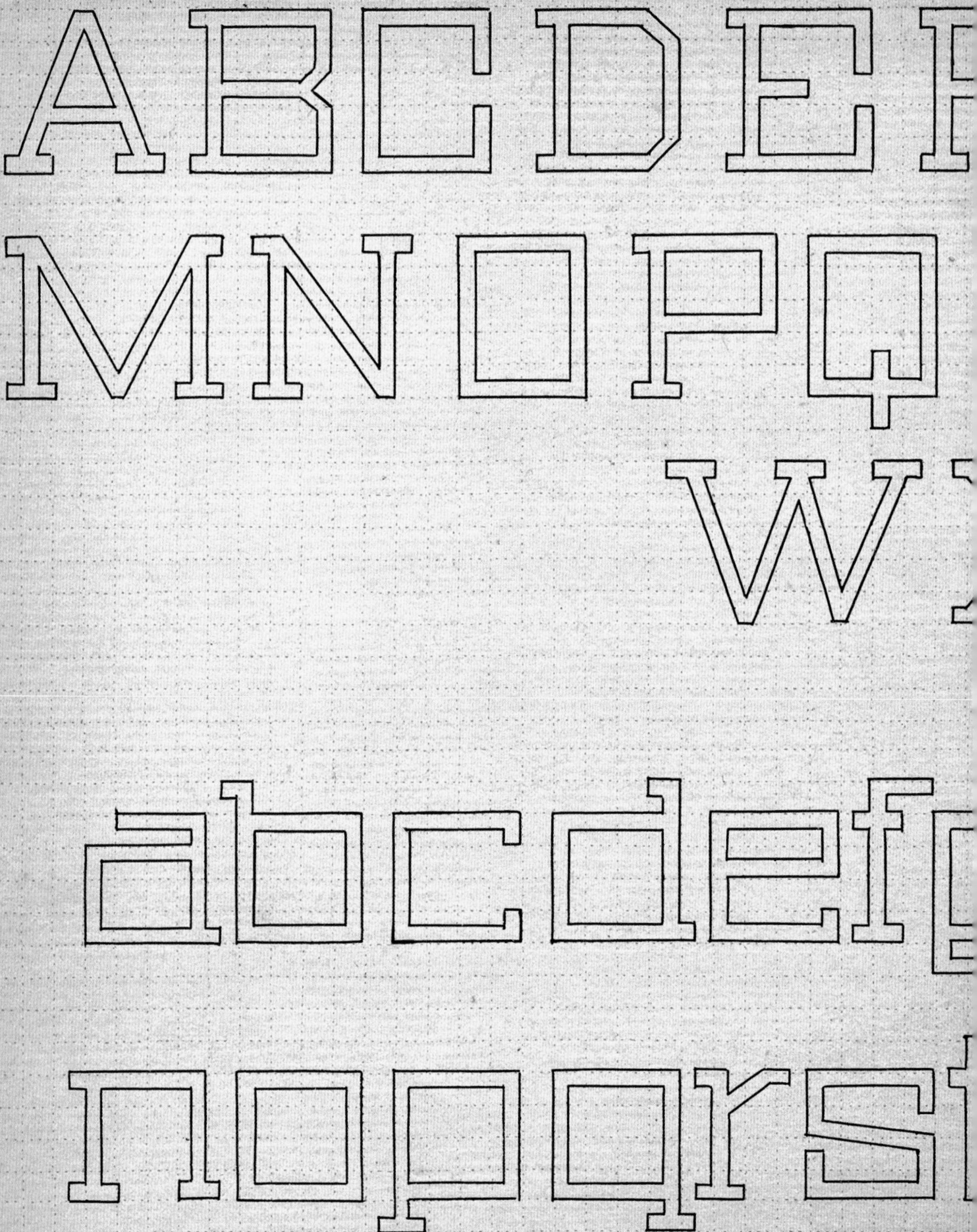

ABCDEF

MNOPQ

W

abcdef

nopqrs

GHIJKL
RSTUVZ
XY

hijklm

UVWXZ

9. *American Jubilee* program cover for the New York World's Fair, designed by Joseph Binder, 1939.

10–11. Styles of gothic and slab serif letters, date unknown.

12–13. Specimen sheet of The Gay Nineties, nineteenth-century ornamented letters, *c.* 1930.

14. Grain sack, Yeager Milling Co., *c.* 1890.

15. Periodical cover for *The Comic Almanac*, 1875.

16–17. Mastheads from weekly and monthly periodicals, 1875–1899.

18–19. Advertising page from German-American newspaper, 1896.

20–21. Merchants' mailing labels and business cards, *c.* 1895.

22–23. Specimen pages from Morgan Press Wood Type catalog, 1965.

24. Ink bottle and label, Thaddeus Davids & Co., 1853.

25. Specimen page from Morgan Press Wood Type catalog, 1965.

26–27. Back and front cover for Burpee's Seeds catalog, 1919.

28–29. Seed packets, Cards Seed Co., *c.* 1920.

30–31. Fruit box labels, 1940s–50s.

32–33. Lettering template from Palm's Patent Transfer Letters, 1885.

34–35. Snellen eye-test chart, 1940s.

36–37. Specimen pages from Morgan Press Wood Type catalog, 1965.

38. Magazine cover for *The American Printer*, August 1931.

39. Brochure, *Letterheads: Then and Now*, 1935.

40. Book jacket for *Inquest*, designed by Arthur Hawkins Jr., 1945.

41. Exhibition catalog cover for *The Textbook*, AIGA, February 1936.

42–43. Bread wrapper, Pullman Loaf, *c.* 1938.

44–45. Subway car advert for Phillies cigars, *c.* 1930.

46–47. Detail of album cover for *Operetta Potpourri*, designed by Alex Steinweiss, 1940s.

48. Book jacket for *Europe In Revolt*, 1942.

49. Periodical cover for *More Business*, designed by Smythe, August 1938.

50–51. Front cover and inside page from *PM* magazine, designed by Leo Rackow, December 1938–January 1939.

52. Operating instruction brochure, Hopalong Cassidy Radio, 1955.

53. Magazine cover for *Western Advertising*, February 1934.

54. Specimen sheet of Stymie Inline Title, American Type Foundry, *c.* 1938.

55. Magazine cover for *The American Printer*, May 1933.

56. Magazine cover for *Advertising Arts*, September 1934.

57. Magazine cover for *Focus*, April 1938.

58. Cookbook cover for *The New Art of Modern Cooking*, *c.* 1938.

59. Catalog cover for *Everything for the Garden*, 1937.

60. Proof of wood type, courtesy of Hamilton Wood Type & Printing Museum, Two Rivers, Wisconsin, USA, 2015.

61. Proof of wood type, courtesy of Ross MacDonald, Brightwork Press, Newtown, Connecticut, USA, 2015.

62–63. Font of shadow letter wood type, courtesy of Ross MacDonald, Brightwork Press, Newtown, Connecticut, USA, 2015.

64–65. Proof of wood type, courtesy of Hamilton Wood Type & Printing Museum, Two Rivers, Wisconsin, USA, 2015.

66–67. Magazine covers for *Town & Country*, January 1937 and November 1935.

68. Magazine cover for *New Pencil Points*, designed by Bernard Rudofsky, January 1943.

69. Magazine cover for *Theatre*, 1948.

70–71. Magazine covers for *New Pencil Points*, Nos 6 & 7, June and July 1943.

72. Specimen sheet of Tower Series, American Type Founders Sales Corporation, *c.* 1935.

73. Advertisement for Sterling Engraving Co., designed by Lester Beall, 1939.

74. Specimen sheet of Cairo, Intertype, *c.* 1935.

75. Book cover for *Norris Dam*, *c.* 1935.

76–77. Specimen sheet of Girder Light and Heavy, Continental Typefounders Association Inc., *c.* 1936.

78–79. Specimen sheet of Rockwell Antique, American Type Founders Company, 1936.

80–81. Specimen sheet of Stymie Bold Condensed, American Type Founders Company, 1938.

82–83. Specimen sheet of Karnak Obelisk, Ludlow Typograph Company, *c.* 1935.

84. Magazine cover for *PM*, designed by Lester Beall, November 1937.

85. Cover for specimen sheet of Beton, Intertype, *c.* 1936.

86–87. Specimen sheets of Beton Bold and Extrabold, Intertype, *c.* 1936.

88. Cover for specimen sheet of City, Continental Typefounders, 1932.

89. Specimen sheet of Playbill, Kurt H. Volk Typography, 1940.

90–91. Pages from specimen book of Memphis Bold Condensed, Luna, and Open, D. Stempel Typefoundry, 1930.

92–93. Specimen sheets of Memphis Light and Stymie Light, American Type Founders Company, *c.* 1935.

94. Book jacket for *The Croquet Player*, 1937.

95. Book jacket for *The Last Puritan*, 1935.

96. Book cover for *Marriage*, 1934.

97. Book cover for *Martha Graham*, designed by Merle Armitage, 1937.

98–99. Specimen sheet of Memphis, American Type Founders Company, *c.* 1935.

100–1. Specimen pages from Morgan Press Wood Type catalog, 1965.

102–3. Show card lettering sample, 1928.

ITALIAN

S LAB SERIFS ARE NOT THE TYPE STYLE THAT FIRST COMES TO mind when thinking about Italy; that would be Bodoni. However, slabs are found on signs, newspapers, magazines, books, food labels, packages, and more. During the late nineteenth and much of the twentieth centuries Italian type shops both large and small filled their type cases with eye-catching expanded and elongated slab serif faces in wood and metal. There was an incredible commercial demand for types that were clear enough to be legible from a reasonable distance yet distinctive enough to have personality. Not all slab serifs produced in Italy possessed these virtues but many satisfied the commercial need to attract attention. Advertisements produced by Campari in the early 1930s to promote their famous herbal cordial employed various slabs, including an angular custom hybrid with a chiseled look (p. 133) and an even heavier alphabet with cut-off serifs (p. 136). Both faces jumped from the black-and-white newsprint page and even overpowered the official logo on the bottle. Although the majority of these extreme serif types fit under the Egyptienne umbrella, many examples draw on the square serif Italienne. This blocky nineteenth-century stylization is characterized by what Jan Tschichold described in *Treasury of Alphabets and Lettering* as "exaggerated horizontal and weak vertical strokes," which make them appear top- and bottom-heavy. "Though not exactly distinguished for legibility," Tschichold added, "this letter, like everything unusual, is very conspicuous."

Slab serifs designed in Italy came from a few leading national foundries, including the Società Augusta in Turin, but not all of those used were originally Italian. The generically titled Tacca Francese (pp. 116–17) was a French import, whereas the faces named after the Piedmont region and the cities of Palermo and Genoa did take root in Italy. The 144 pt. face under the rubric Tacca Italiana is more eccentric, with exaggerated curvilinear serifs giving the upper- and lower-case specimen an optically fluid quality. In contrast, the 120 pt. Tacca Italiana is stark, heavy, and stern. Some letters, like in the words "Cornici" and "Libreria" (pp. 124–25), project a no-nonsense functionalism, while the inventively quirky, irregular letters for "Cappelleria" (pp. 128–29) and "Fiori" (p. 130) play a more comedic role.

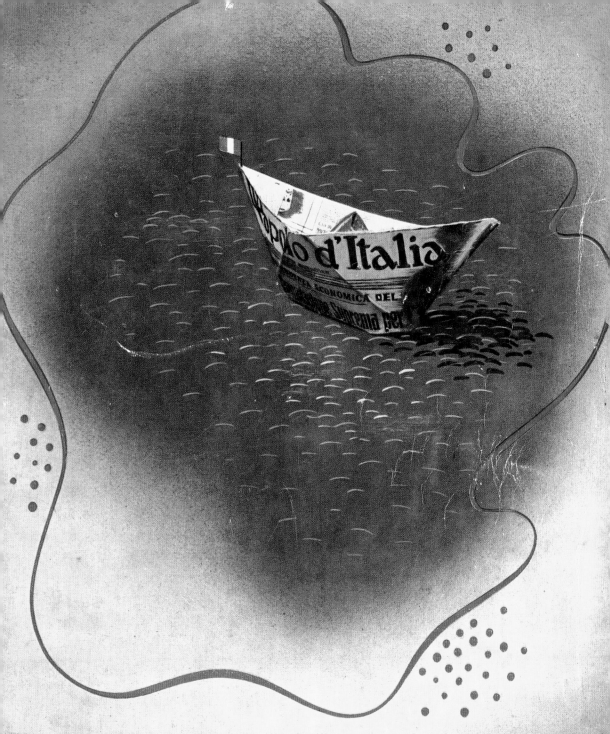

NO III · N. 25-30 · LUGLIO-DICEMBRE 1939-XVIII · SPEDIZ. ABB. POST. · LIRE 45

LA PUBBLICITÀ D'ITALIA

Anno I.   (Conto Corrente colla Posta)   **Milano**

.... Due e due fanno quattro

LA PROPO

ORGANO DELL'ASSOCIAZIONI

Dirigere corrispondenze e manoscritti all'Avv. L

Donahà la D. D. niroand   basterà e sarà

SABATO 7 SETTEMBRE 1839

CONDIZIONI PER L'ASSOCIAZIONE.

L'OMNIBUS.

FOGLIO.. GR 5
SEMESTRE....................................... 1 20

108

I pagamenti sempre anticipati.

IN

RELIGIONE. *Progressi della fede.* -
RATURA. — *Schiller e Goethe.* — COS
GRAFICA. *Ore d'ozio - Cenno sugli educ
Nina siciliana, o nina di Dante.* — ART
d'uva. — VARIETA. *La giornata d'un
perla Ottico-pittorica.* — CORRISPONDEN
TEATRI. — NOTIZIE TEATRALI. — AL

Luglio 1919   Un numero Cent. 10   N. 8.

Esce ogni settimana

RZIONALE

ROPORZIONALISTA MILANESE ——

I DEGLI OCCHI - Milano - Piazza Borromeo, 10

e anche troppo   socialista e cattolico per squali-

ANNO SETTIMO N. 19

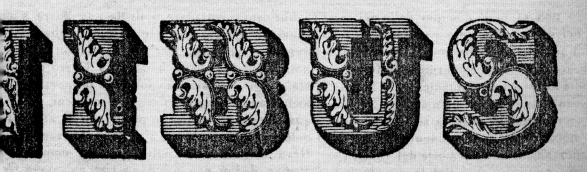

E

CA. *Il sistema di Galileo.* — LETTE-
Il sangiovannaro. — RIVISTA BIBLIO-
egina isabella Borbone. — BIOGRAFIA.
ESTIERI. *nuova macchina da spremere*
— INVENZIONI E SCOPERTE. *Sco-*
ACCADEMIE. — COSE D.VERSE. —
— SCIARADA. —

CONDIZIONI PER L'ASSOCIAZIONE

PITTORESCO E LETTERARIO

UN SEMESTRE . . . . 2. (

UN'ANN•PA . . . . 4. (

109

pagamenti sempre anticipati

GIOVANI

NEGOZIANT

Deposito e Vendita d

IN M

Barbieri

Modena li

...I POZZI

...DI CARTA

...Oggetti di Cancelleri...

...DENA

ENGLISH BAKER

FORNITORI

DIEU ET MON DROIT

HONI SOIT QUI MAL PENSE

G. COL

BISCUIT MA

RO

Via del Bab

WHOLESALE GROCER

REALCASA

LUCCI

UFACTURER

ME

no N° 94

113

Bam

'Tacca Francese

Ermafrodito

bina

Club Automobilistico

Benedetto Cairoli

MARCATI

PREMIATA FABBRICA LIQUORI

PIETRO MARCATI

TREVISO

FUORI PORTA MAZZINI

PREMIATA AL CONCORSO INDUS.
DEL R. ISTITUTO VENETO DI SCIENZE
LETTERE ED ARTI - VENEZIA 1901

SLIWOWITZ ITALIANO

SPECIALITÀ
DELLA DITTA
P. MARCATI

119

Tacca Italiana

Bastion

Tacca Italiana

Hecho

Tacca Italiana

Male

Merveilleuse Instruction

Internazionalista

Recomendaciones

Tacca Italiana

ESAURIMENTO

Tacca Italiana

HERMOSAS

Tacca Italiana

COUSINE

BENDE

LOUIS

RISE

BOTTEGA

APPARECCHI
D'ILLUMINAZIONE

F. CE

LIBR

CARTO

COR

DELLA LUCE

CONI

MATERIALE
ELETTRICO

ERIA

LERIA

NICI

FORNO A

AMEDEO

OST

OLI

VAPORE

GIUSTI

ERIA

VINI

CAPPE

dell

LATTE PANNA
E ANTICA
BURRO
DI DARIO

non c'è Natale senza "Cordiale"

ritorna il presepe....

1001

.... ritorna la cara consuetudine, che alla grande solennità cristiana aggiunge un incomparabile riflesso di festosa intimità. Come, nella festosa intimità delle famiglie, aggiunge luce al sorriso un sorso di CORDIAL CAMPARI tradizione antica e nuova d'Italia

CORDIAL CAMPARI

il liquore delle famiglie italiane

CAMPARI

Cordial Campari

133

CAM
CORDI

IN VIAGG

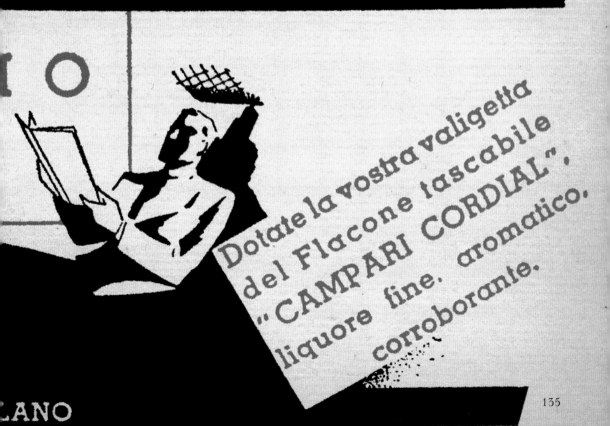

PARI
AL Liquor

o

Dotate la vostra valigetta
del Flacone tascabile
"CAMPARI CORDIAL",
liquore fine. aromatico,
corroborante.

LANO

UNA DELIZIOSA ABITUDINE

CAMPARI

Bitter

G. Campari

Prendere un aperitivo prima di pranzo è ormai nell'abitudine di tutti. Prendere un BITTER CAMPARI, l'aperitivo per eccellenza, è la deliziosa abitudine di tutte le persone di buon gusto.

Bitter CAMPARI

l'aperitivo

IL VESTITO ANTINEUTRALE

Manifesto futurista

> Glorifichiamo la guerra,
> sola igiene del mondo.
> **MARINETTI.**
> *(1° Manifesto del Futurismo - 20 Febbraio 1909)*

Viva Asinari di Bernezzo!
MARINETTI.
(1ª Serata futurista - Teatro Lirico, Milano, Febbraio 1910)

L'umanità si vestì sempre di **quiete**, di **paura**, di **cautela** o d'**indecisione**, portò sempre il lutto, o il piviale, o il mantello. Il corpo dell'uomo fu sempre diminuito da sfumature e da tinte **neutre**, avvilito dal nero, soffocato da cinture, imprigionato da panneggiamenti.

Fino ad oggi gli uomini usarono abiti di colori e forme statiche, cioè drappeggiati, solenni, gravi, incomodi e sacerdotali. Erano espressioni di timidezza, di malinconia e di **schiavitù**, negazione della vita muscolare, che soffocava in un passatismo anti-igienico di stoffe troppo pesanti e di mezze tinte tediose, effeminate o decadenti. Tonalità e ritmi di **pace desolante**, funeraria e deprimente.

OGGI vogliamo abolire:

1. — Tutte le tinte **neutre**, « carine », sbiadite, *fantasia*, semioscure e umilianti.

2. — Tutte le tinte e le foggie pedanti, professorali e teutoniche. I disegni a righe, a quadretti, a **puntini diplomatici.**

3. — I vestiti da lutto, nemmeno adatti per i becchini. Le morti eroiche non devono essere compiante, ma ricordate con vestiti rossi.

4. — L'equilibrio **mediocrista**, il cosidetto buon gusto e la cosidetta armonia di tinte e di forme, che frenano gli entusiasmi e rallentano il passo.

5. — La simmetria nel taglio, le linee **statiche**, che stancano, deprimono, contristano, legano i muscoli; l'uniformità di goffi risvolti e tutte le cincischiature. I bottoni inutili. I colletti e i polsini inamidati.

Noi futuristi vogliamo liberare la nostra razza da ogni **neutralità**, dall'indecisione paurosa e quietista, dal pessimismo negatore e dall'inerzia

Vestito bianco - rosso - verde
del parolibero futurista Marinetti. *(Mattino)*

MINEO

FORNITORI DELLA CASA
DI SAR IL PRINCIPE DI PIEMONTE

MED D'ORO TORINO 1911

MEDAGLIA D'ORO TORINO 1911

MED D'ORO GENOVA 1914

MARSALA

GARIBALDI DOLCE G.D.

GIACOMO MINEO & FIGLI

ANTICA CASA FONDATA NEL 1862

MARSALA

CURATOLO & C

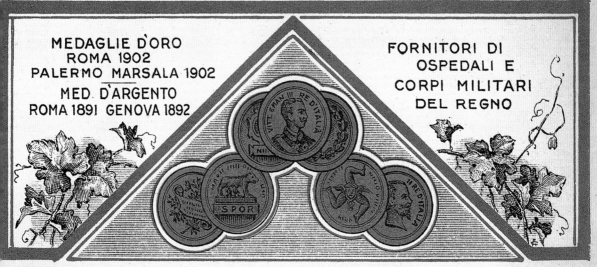

MEDAGLIE D'ORO
ROMA 1902
PALERMO MARSALA 1902

MED. D'ARGENTO
ROMA 1891 GENOVA 1892

FORNITORI DI
OSPEDALI E
CORPI MILITARI
DEL REGNO

MARSALA

GARIBALDI G.D.

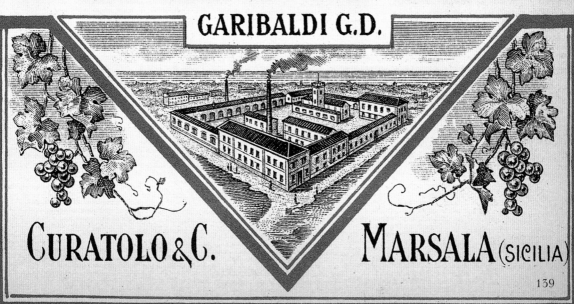

CURATOLO & C.   MARSALA (SICILIA)

139

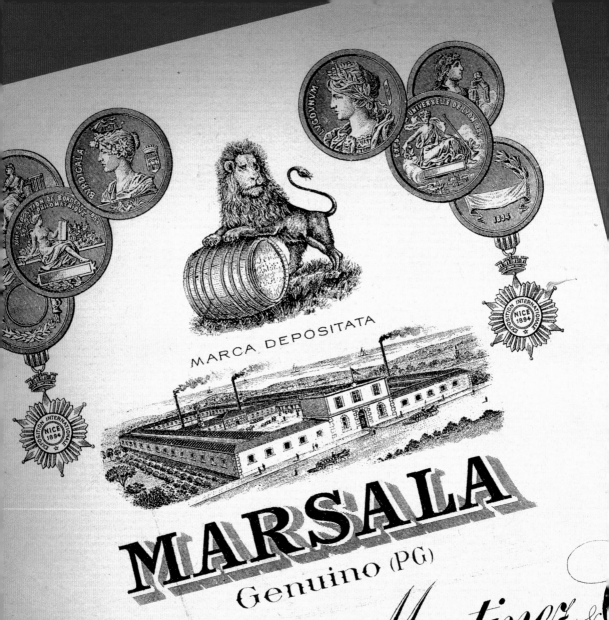

MARCA DEPOSITATA

MARSALA

Genuino (PG)

Domenico Florio Martinez &

MARSAL

RICHTER & C.º NAPOLI.

CALAMIA

TORINO 1892 · FILADELFIA 1914 · MILANO 1900 · PARIGI 1894 · PALERMO 1911 · BOLOGNA 1890 · NAPOLI 1908 · MILANO 1906

MARCA DI FABBRICA

MARSALA

I. P.
ITALIA EXTRA

particolare del Premiato Stab. Enologico

Salvatore Calamia

MARSALA

RIVISTA TURISTICA MENSILE DELL'ENIT
E DELLE FERROVIE DELLO STATO

Anno VI - Numero 4 - Febbraio 1938 - XVI

ITALIA

RIVISTA TURISTICA MENSILE DELL'ENIT
E DELLE FERROVIE DELLO STATO

Anno VII · Numero 5 · Marzo 1939 · XVII

ITALIA

RIVISTA TURISTICA MENSILE DELL'ENIT
E DELLE FERROVIE DELLO STATO

Anno VI - Numero 7 - Maggio 1938 - XVI

ITALIA

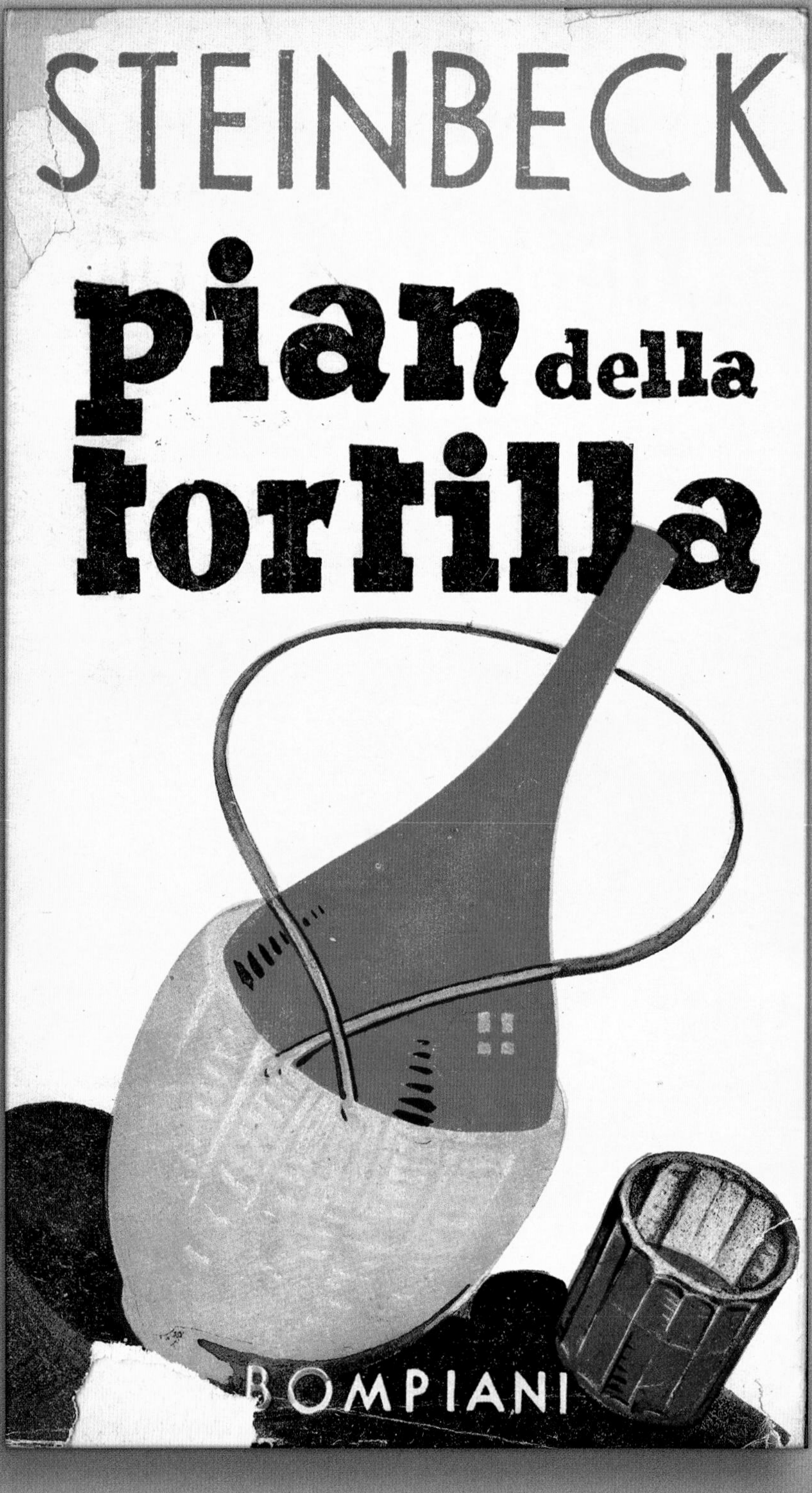
STEINBECK
pian della tortilla
BOMPIANI

DOCUMENTI DI "TEMPO NOSTRO,, N. 2

GIUSEPPE MARANINI

UTOPIA
DOPO LA RIVOLUZIONE

TEMPO NOSTRO

editrice cooperativa - Roma - 1945

ANNO XV - N. 315
1º Ottobre 1939-XVII

il dramma

quindicinale di commedie di grande successo diretto da **lucio ridenti**

Marcello Gio[r...]

SOCIETÀ EDITRICE
TORINESE - TORINO

Lire **1.50**
SPEDIZIONE IN ABBONAMENTO POSTALE

ANNO XIX - N. 401
1° Maggio 1943-XXI

Lire 2,50

SOCIETÀ EDITRICE
TORINESE - TORINO
SPEDIZIONE IN ABBONAMEN-
TO POSTALE (Secondo Gruppo)

il dramma

quindicinale di commedie di grande successo diretto da **lucio ridenti**

ina De Filippo

ONORATO
43

PREMIO NOBEL: tre atti di HYALMAR BERGMAN

IMA DI COLAZIONE: un atto di EUGENIO O' NEIL

L'UOMO CHE RIDE

SETTIMANALE SATIRICO-POLITICO

ROMA - 9 DICEMBRE 1945 - ANNO I° - N. 1 | ESCE LA DOMENICA | REDAZ. AMMIN.: ROMA - VIA A. FARNESE, 11

Lire 10

Fuori Roma L. 1.

Spedizione in abbonamento postale

L'unico che sarà veramente epurato

Siete accusato d'aver scritto «De Monarchia» d'aver collaborato col tedesco, d'aver iniziato la campagna razzista con i noti versi della «Commedia» e di essere andato al nord al servizio del Signore di Verona

(Apoll)

ANNO VII · N. 31 · 2 AGOSTO 1951 ★ SETTIMANALE DI POLITICA ATTUALITÀ E CULTURA ★ SPED. ABB. POST. GR. II ★ LIRE SESSANTA

OGGI

Alle pagine 20-25

MUSSOLINI E IL FASCISMO IN FOTOGRAFIE

IDA LUPINO, LA DONNA

PIÙ INTRAPRENDENTE

DEL CINEMA AMERICANO

Molti spettatori usano considerare le "dive" di Hollywood solo come vuote bambole, prive di vere qualità personali, ove se ne escluda la bellezza. Ida Lupino, già attrice celebre, ha voluto smentire questo luogo comune: da due anni ella fa parlare di sé come regista, produttrice, scrittrice, truccatrice, fotografa. Ha già diretto tre film (prima e unica donna-regista di Hollywood) ed altri si appresta a mettere in cantiere per conto della società produttrice "Filmakers", di cui ella è proprietaria insieme al marito Collier Young. Nata a Londra nel 1918, durante un'incursione aerea tedesca, Ida Lupino è figlia di un noto attore drammatico inglese, Stanley Lupino, discendente a sua volta da una famiglia di acrobati e ballerini di origine italiana. Ancora bambina Ida sosteneva già parti di primo piano; a 15 anni andò ad Hollywood, dove divenne in breve famosa per una serie di interpretazioni in film di successo. Convinta, però, di avere ben altre possibilità artistiche, ella accettò la proposta del secondo marito, il produttore Collier Young (il primo, da cui divorziò, fu l'attore Lous Hayward), di unirsi a lui in società per produrre e dirigere film. Oggi si dice che la Lupino conosca la macchina da presa megio di tutti i maggiori registi: tuttavia le nuove molteplici attività non hanno portato fortuna alla sua vita privata. Ida si è infatti sparata anche da Young, pur restando sua socia in affari (*Vedere nell'interno alle pag. 14-15*).

G.ᴰ HÔTEL BAG
FLO

A INTERNAZIONALE - LEGNAGO (ITALIE)

CARACTÈRES EN BOIS
MEUBLES D'IMPRIMERIE

Serie PIEMONTE

ma

Cic. 8 - Classe A-h

oderna

Cic. 30 - Classe D-g

Bul

Ze

Ri

Buttero

ARATO

Mondo

Cic. 12 - Classe A-o

Caroline

Cic. 18 - Classe B-f

Dans

Cic. 20 - Classe B-i

Bâle

Born

CAIO

nida!

MARIO

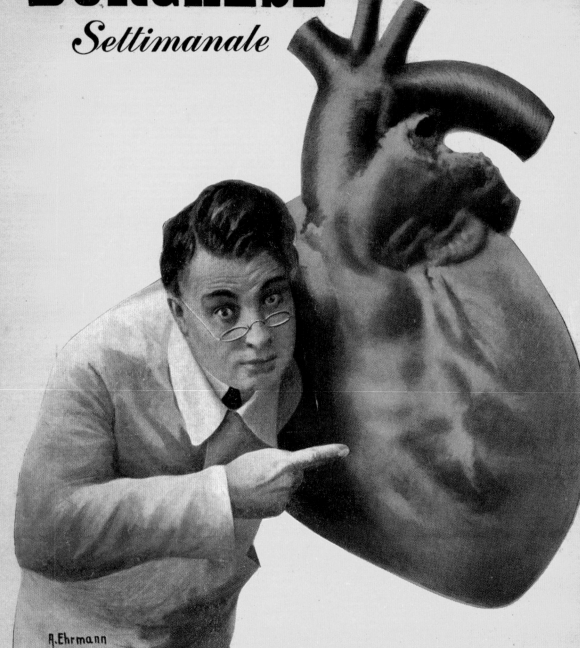

il BORGHESE

Settimanale

A. Ehrmann

numero 17

Longanesi & C.

il BORGHESE
numero 44

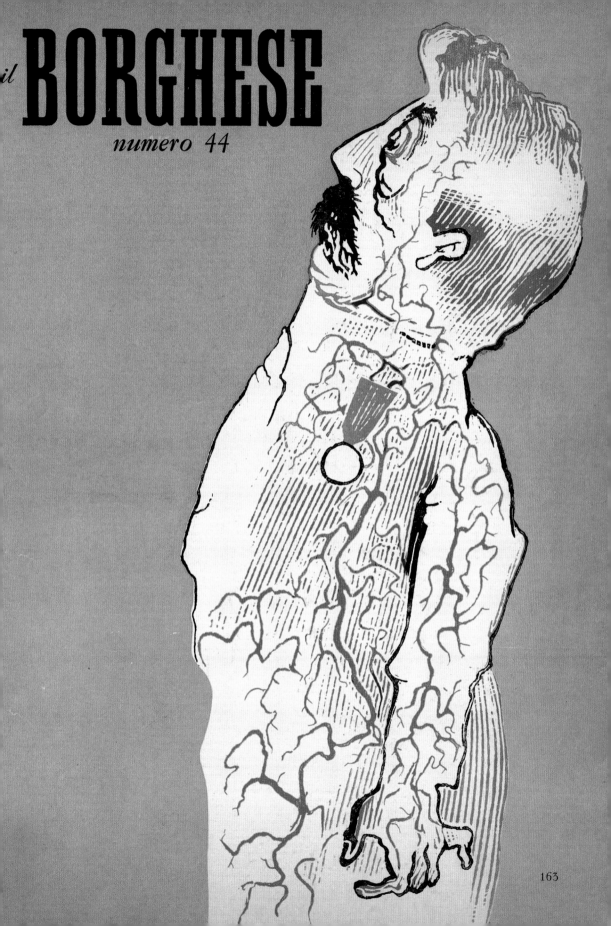

FILA
la matita Italiana di qualità
FABBRICA ITALIANA LAPIS E AFFINI-FIRENZE

la mat

ILA

a Italiana di qualità

FILA ✳ TELEFONO ✳ 401/Nº 2

LA ✳ TELEFONO ✳ 401/Nº 2

FILA ✳ TELEFONO ✳ 401/Nº 2

FILA ✳ TELEFONO ✳ 401/Nº 2

LA ✳ TELEFONO ✳ 401/Nº 2

INCHIOSTRI · MATITE ·

Matita
Copiativa
Finissima

IS

DITTA A. BO

FIM·TORINO·FABBRIC

RONI

★

Z

MARCA DI FABBRICA

bruciare nelle stanze

ntro le Zanzare

PASTELLI
GIOTTO
A COLORI

N. 1366 con 6 Pastelli
N. 1372 con 12 Pastelli

FILA
FIRENZE

107. Magazine cover for *La Pubblicità d'Italia*, July–December 1939.

108–9. Periodical mastheads for *La Proporzionale*, July 15, 1919, and *L'Omnibus*, September 7, 1859.

110–11. Billhead, Giovanni Pozzi, 1892.

112–13. Billhead, G. Colalucci, 1895.

114–15. Billhead, Fratelli Monti, 1895.

116–17. Specimens of Tacca Francese, *Campionario caratteri e fregi tipografici: parti prima, seconda, e terza*, Società Anonima ditta Nebiolo & Comp., 1920.

118–19. Aperitivo labels, 1910s.

120–23. Specimens of Caratteri per Avvisi, *Campionario caratteri e fregi tipografici: parti prima, seconda, e terza*, Società Anonima ditta Nebiolo & Comp., 1920.

124–31. Italian signs, photographed by Louise Fili, 2010–2013.

132–36. Advertisements for Campari, 1931–1933.

137. Front page of *Il Vestito Antineutrale: Manifesto futurista*, designed by Giacomo Balla, 1914.

138–41. Marsala wine labels, *c.* 1930.

142–45. Magazine covers for *Italia*, 1938–1939.

146. Book jacket for the Italian edition of *Tortilla Flat*, designed by Venna, 1941.

147. Pamphlet cover for *Utopia dopo La Rivoluzione*, 1945.

148–49. Magazine covers for *Il Dramma*, October 1939 and March 1943.

150. Periodical cover for *L'Uomo Che Ride*, December 9, 1945.

151. Periodical cover for *Oggi*, August 2, 1951.

152–53. Luggage label, Grand Hôtel Baglioni & Palace, *c.* 1930.

154–55. Wood type specimen page of Serie Piemonte, *S. A. Xilografia Internazionale–Catalogo caratteri di Legno, Parte I-II-III*, *c.* 1930.

156–57. Wood type specimen page of Serie Palermo, *S. A. Xilografia Internazionale–Catalogo caratteri di Legno, Parte I-II-III*, *c.* 1930.

158. Wood type specimen page of Serie Piemonte, *S. A. Xilografia Internazionale–Catalogo caratteri di Legno, Parte I-II-III*, *c.* 1930.

159. Wood type specimen page of Serie Genova, *S. A. Xilografia Internazionale–Catalogo caratteri di Legno, Parte I-II-III*, *c.* 1930.

160–61. Wood type specimen page from *S. A. Xilografia Internazionale–Catalogo caratteri di Legno, Parte I-II-III*, *c.* 1930.

162–63. Magazine covers for *Il Borghese*, nos 17 and 44, dates unknown.

164–65. Display card, Fila pencil, *c.* 1940.

166–67. Pencil box, Doris, date unknown, and fungicide package, Zampironi, *c.* 1935.

168. Crayon box, Pastelli Giotto a Colori (Fila), *c.* 1928.

FRENCH

SOMEWHERE EXISTS A HELLBOX OF ANCIENT TYPEFACES THAT have defined French graphic design at various times during its grand history, but slab serifs hold a unique status. Used for newspaper mastheads, business logos, and product packages, slab serifs' ubiquity is connected to their Napoleonic roots. The Emperor Bonaparte did not actually design typefaces any more than he personally prepared the delicious *mille-feuille* pastry known as the Napoleon. However, the peripatetic French adventurer led an expedition into Egypt and subsequently commissioned the *Description de l'Égypte*, an ambitious multi-volume study of the country and its natural history written by scholars who accompanied the army's campaign. Its publication prompted a popular fascination with all things Egyptian and inspired the nomenclature of a novel type genre, Egyptiennes.

Slab serifs were actually born in Britain, then exported everywhere type was used. Their beefy black forms derived from shifts in aesthetics triggered by the Industrial Revolution. Influenced by commercial demands, French type foundries such as Deberny et Peignot, Fonderie Warnery, Fonderie Typographique Gustave Mayeur and Allainguillaume & Cie. produced massive stocks of Eyptiennes, which they advertised as "for the frequent employ of black letters in ordinary texts." Many metal fonts were sold to print shops by the kilo, used for body and headlines (p. 210), while others, such as Les Italiennes Allongées (p. 175), were specially cut for display. Additionally, ornamental baroque versions of slab serifs were engraved or etched to provide printers with page enhancements. These included letters with bifurcated serifs, known as Fantaisies (p. 189). Fonderie Warnery's eclectic Egyptiennes Dentelles, so called for their serrated bodies, were not appropriate for all jobs, but were visually appealing when used judiciously. There was also a surge in hand-drawn slabs that were riddled with imperfections (p. 226). The type expert Douglas C. McMurtrie wrote: "The design of square serifs had been encouraged, I believe, by the wide use which French modern typographers had been making of typewriter types … because in the square serif family tree, typewriter type is at least a legitimate second cousin." It was the sheer diversity of the slab genre that accounted for the longevity of the style. *Vive la différence*!

Les Voyageurs Modernes...

connaissent la valeur du temps. Ils savent que les voitures RENAULT et les avions CAUDRON-RENAULT leur permettent les plus grands déplacements avec le maximum d'agrément et de sécurité, le plus grand confort et la plus stricte économie. Les voyageurs modernes choisissent toujours les solutions de l'expérience au service de la vie moderne:

RENAULT

FONDERIE MAYEUR

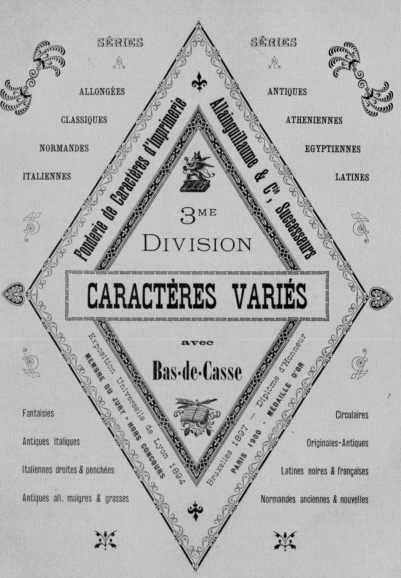

SÉRIES   SÉRIES

ALLONGÉES   ANTIQUES

CLASSIQUES   ATHENIENNES

NORMANDES   EGYPTIENNES

ITALIENNES   LATINES

Fonderie de Caractères d'Imprimerie
Allainguillaume & Cie, Successeurs

3ME DIVISION

CARACTÈRES VARIÉS

avec

Bas-de-Casse

Exposition Universelle de Lyon 1894
MEMBRE DU JURY - HORS CONCOURS

Bruxelles 1897 - Diplôme d'Honneur
PARIS 1900 - MÉDAILLE D'OR

Fantaisies   Circulaires

Antiques italiques   Originales-Antiques

Italiennes droites & penchées   Latines noires & françaises

Antiques all. maigres & grasses   Normandes anciennes & nouvelles

PARIS ⚜ Rue du Montparnasse, 21 ⚜ PARIS

ÉGYPTIENNES CARRÉES (avec bas de casse)

créées pour les emplois fréquents de types noirs intercalés dans les textes ordinaires,
par la Fonderie MAYEUR

ALLAINGUILLAUME & Cie Successeurs ❧ 21, rue du Montparnasse ❧ PARIS

PRIX PAR KILO

1221 — Corps 5 — 16 fr.

Les Éléments de Géométrie avec les Exercices et Problèmes d'Algèbre

100 a 50 A : env. 2 k. 15

LES SERVICES GÉOGRAPHIQUES DES ARMÉES 1234567890

1222 — Corps 6 — 10 fr.

Publication du Nouveau Recueil des Comédies et des Drames

100 a 50 A : env. 2 k. 85

MARSEILLE, TOULOUSE, LYON, NICE 1234567890

1223 — Corps 7 — 8 fr.

Le Rat et la Belette, une des fables de La Fontaine, est

100 a 50 A : env. 3 k. 80

LA MARCHE DES CHAUFFEURS 1234567890

1224 — Corps 8 — 7 fr.

Dessins d'ornementation moderne en deux couleurs

100 a 50 A : env. 4 k. 80

MAISON & DÉPENDANCES 1234567890

1225 — Corps 9 — 6 fr.

Merveilles Naturelles du Globe Terrestre

100 a 50 A : env. 6 k. 15

RÉOUVERTURE LE 25 MARS 1904

1226 — Corps 10 — 5 fr. 50

Grande Collection de Timbres-Poste

50 a 25 A : env. 3 k. 90

VENTE DE 18,240 VOLUMES

1227 — Corps 12 — 5 fr.

Les Récréations des Écoliers

50 a 25 A : env. 6 k. 40

L'ANNÉE 1875 A BREST

1228 — Corps 16 — 4 fr. 75

La Marine Française

50 a 25 A : env. 11 k.

GUERRE DE 1870

1229 — Corps 18 - 4 fr. 50 — Corps 20 - 4 fr. 25

Carnaval de Venise

50 a 25 A : env. 14 k. 80

HUMILIATIONS

807 — Corps 24 — * 3 fr. 75

Histoire naturelle

50 a 25 A : env. 18 k.

LA MASCOTTE

808 — Corps 28 — * 3 fr. 50

Mairie de Caen

25 a 12 A : env. 13 k.

NORMANDIE

809 — Corps 36 — * 3 fr. 25

Préfectures

25 a 12 A : env. 21 k.

ESCRIME

810 — Corps 48 — * 3 fr.

Ramiers

15 a 7 A : env. 25 k.

NIMES

871 — Corps 60 — 3 fr.

Larme

10 a 5 A : env. 27 k.

ECHO

1083 — Corps 72 — 2 fr. 90

Beau

10 a 5 A : env. 40 k.

MER

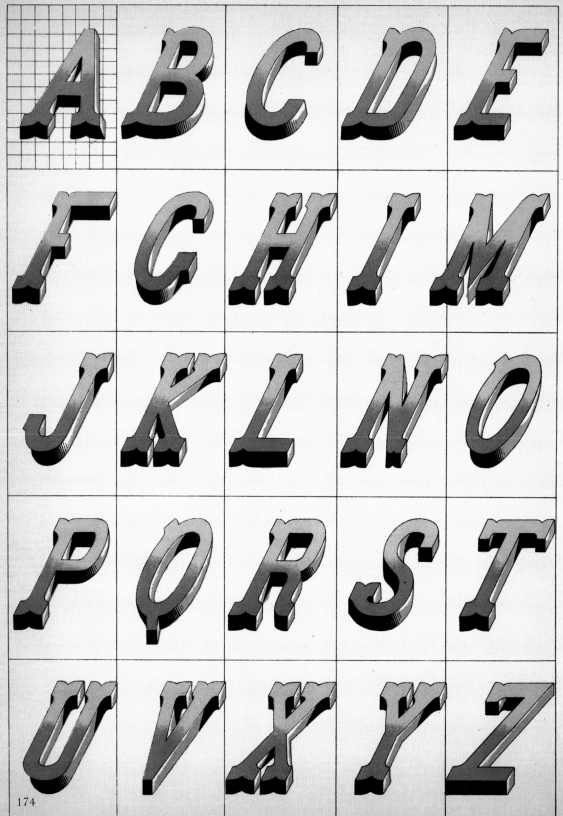

Emile Thézard et fils Éditeurs à Dourdan.(S.-&-O.)

IMP. LEVERT et Cie PARIS

Les Italiennes Allongées

50 A
env. 2 kil. 50 — 402 — Corps 12 — *7 fr. le kilo

403 — Corps 16 — *6 fr. 50
30 A
env. 3 kil. 10

PROMENADE AU BOIS 123456

GRANDES ŒUVRES 789

277 — Corps 20 — *6 fr.

25 A : env. 3 kil.

LA GÉOGRAPHIE UNIVERSELLE 1324567890

404 — Corps 28 — *5 fr. 50

20 A : env. 5 kil. 40

BOULEVARD DE LA MADELEINE 368

321 — Corps 36 — *5 fr.

10 A : env. 6 kil.

GRENADIERS DE FRANCE

405 — Corps 40 — ** 5 fr.

10 A : env. 7 kil. 50

MÉMORIAL DE ROUEN

372 — Corps 48 — ** 4 fr. 50

10 A : env. 10 kil.

CONQUÊTES 237

373 — Corps 56 — ** 4 fr.

7 A : env. 10 kil. 30

MONTMARTRE

295 — Corps 72 — ** 3 fr. 25

5 A : env. 10 kil.

24 TITRES

175

Nᵒ 14257 — 3 Germinal An 117.

CINQ CENTIM

Fondateur :
AUGUSTE VACQUERIE

ABONNEMENTS

	Un mois	Trois mois	Six mois	Un an
Paris..........	2 fr.	5 fr.	9 fr.	18 fr.
Départements	2 —	6 —	11 —	20 —
Union Postale	3 —	9 —	16 —	32 —

Adresser toutes les Communications au Directeur

LE RA

ADMINISTRATION & RÉDACTION : 53, rue du Château-d'Eau : Téléphone 438-1

TRIBUNE LIBRE

Nᵒ 72 — 2ᵉ ANNÉE.   10

Dans la Seine, où, au ballottage,

LE BON

Journal Humori

4ᵐᵉ ANNÉE. 2ᵐᵉ Série. — Nᵒ 92. Le numéro (15

MONDE PI

Voyages. — Découvertes, Explorations. — Aventures de
Récits de Chasse et de Pêche. — Curiosi

176

E NUMERO   Mercredi 24 Mars 1909. — N° 14257

PPEL

Fondateur :
AUGUSTE VACQUERIE

ANNONCES
MM. LAGRANGE, CERF & C⁞
6, Place de la Bourse
et aux BUREAUX DU JOURNAL

Adresser Lettres et Mandats au Directeur

9 heures du soir à 2 heures du matin, 123, rue Montmartre : Téléphone 143-93

re, secrétaire de mairie, amis intimes,
sans exception, tous font élever leurs

.times

A LA CHAMBRE

30 Mars 1901

VIVANT

que de la Famille

de texte) : **15 cent.**   DIMANCHE 5 OCTOBRE 1884.

TORESQUE

e et de Mer. — Mœurs et Coutumes des différents Peuples
ographiques et scientifiques. — Romans.

177

Le Petit Journal

ADMINISTRATION
61, RUE LAFAYETTE, 61

Les manuscrits ne sont pas rendus

On s'abonne sans frais
dans tous les bureaux de poste

5 CENT. SUPPLÉMENT ILLUSTRÉ **5** CENT.

24me Année ✱✱ Numéro 1.174

DIMANCHE 18 MAI 1913

ABONNEMENTS

	SIX MOIS	UN AN
SEINE et SEINE-ET-OISE..	2 fr.	3 fr. 50
DÉPARTEMENTS.............	2 fr.	4 fr.
ÉTRANGER	2 50	5 fr. 5

HONNEUR A JEANNE D'ARC !

Un régiment passant devant la maison de Domrémy
salue le logis natal de l'héroïne

2e SÉRIE — No 1　　　16, RUE DU CROISSANT, 16　　　PRIX : 5 CENT.

LE MONDE COMIQUE

Aux Bureaux de l'Éclipse et de la Lanterne de Boquillon

UN NUMÉRO : 5 CENTIMES. — Un an, Paris, 4 fr. — Départements, 5 fr — UN NUMÉRO : 5 CENTIMES

LES RUINES DE PARIS, par HADOL.

Arthur va dans le monde.

2ᵉ Année. Nº 5   Dimanche 26 Octobre 1879   PRIX : 10 cent.

Le Père Gérard

GAZETTE NATIONALE DES COMMUNES

ABONNEMENT		ADMINISTRATION
Un an............ 6 f.	Rédigé par E. BOURSIN. — Illustré par Léonce PETIT	ET RÉDACTION ⑨
Six mois .... 3 f.	Bureaux à Paris, 15, rue Malebranche.	à St-Germain-en-Laye

LETTRE DE MATHURIN AU PÈRE GÉRARD SUR LES AFFAIRES DE LA POLITIQUE

DROLERIES

CONTEMPORAINES

ALBUM

DE 60 CARICATURES

10ᵉ ANNÉE. — Nᵒ 6.

Édition de Paris : 5 centimes.

9 FÉVRIER 1900

RÉDACTION & ADMINISTRATION
33, rue de Provence, Paris

ABONNEMENTS :
GIL BLAS Quotidien

Trois mois | Paris........ 13 fr. 50
| Départements 15 fr. »

Prix du Numéro :
PARIS ET PROVINCE : O fr. 15

RÉDACTION & ADMINISTRATION
33, rue de Provence, Paris

Toute la Correspondance doit être
adressée à l'Administrateur

ABONNEMENTS :
GIL BLAS Illustré

Paris et Départ. Étranger.
Trois mois...... 1 fr 50 2 fr. 50
Six mois...... 3 — » 5 — »
Un an...... 6 — » 10 — »

GIL BLAS

ILLUSTRÉ HEBDOMADAIRE

*Amuser les gens qui passent, leur plaire aujourd'hui et recommencer
le lendemain. — J. Janin, préface de GIL BLAS.*

POUR CETTE CAUSE SAINTE, par GUSTAVE GUICHES

TROISIÈME ANNÉE — N° 59-60 — SAMEDI 2 AVRIL 1898

15 CENT

15 CENT

LA NOUVELLE
REVUE PARISIENNE

Littéraire, Artistique, Hebdomadaire

RÉDACTION & ADMINISTRATION : RUE DE BEAUNE, 14, PARIS

BUREAUX A BRUXELLES, 36, RUE LONGUE-VIE

CE NUMÉRO CONTIENT

Des ARTICLES de ACH. SEGARD, ED. PONTIÉ, L. MOCHE, AXEL, etc.
(18 colonnes de texte)
2 poses de la photographie de Mlle Julienne Bellanger.

PRIX DU NUMÉRO :

FRANCE & BELGIQUE
15 CENT.

Étranger : 25 cent.

EN VENTE :

dans tous les kiosques de Paris ;
dans toutes les Villes des Départements ;
dans les principales villes étran-
gères.

Photographie REUTLINGER, Paris.

MULNIER

ABONNEMENT ANNUEL :

FRANCE : 6 FRANC
Étranger : 8 francs

ON S'ABONNE :

au Bureau du journal ;
dans tous les bureaux de poste ;
chez tous les libraires.

Bureaux ouverts de 2 à 3 heur

Julienne BELLANGER

DU CHATELET

SOCIÉTÉ ANONYME DE CONSTRUCTION

DE LA

VILLETTE

USINES DE LIVRY (SEINE & OISE)

GRANDE FABRIQUE

DE

MOULURES POUR BATIMENTS · MEUBLES DE CUISINE · BOITES ET MALLES DE VOYAGE

ALBUM

DE MOULURES POUR BÂTIMENTS · CHÈNE & SAPIN

> ÉDITION 1880 <

Adresser les Commandes & Correspondances

à Mr MOUTON, Représt

AUX MAGASINS & DÉPÔTS

29, Rue Château-Landon

PARIS

8297   Imp GRAND, 14, Rue du Faubourg St Martin, PARIS.

ET POUR LES ROULER MIEUX ENCORE
RIZLA+
BLEU A LETTRES D'OR
CASSIS
LITH V. PALART PARIS
N° 484

NICOLAS
les produits de qualité

Jus D'ORANGE

Confitures
NICOLAS
MARQUE DÉPOSÉE
TUNIS

SOCIÉTÉ DES FERMES FRANÇAISES
FF

EGYPTIENNES ALLONGÉES

Corps 9.

12345   LES JOLIS BOUQUETS DE VIOLETTES SONT DES   67890

Corps 12.

67890   LES CAMPAGNES MALHEUREUSES AUX   12345

Corps 16.

12345   LE PALAIS DES BEAUX-ARTS   67890

Corps 18.

67890   EIFFEL, INGÉNIEUR &   12345

Corps 20.

12345 LE PETIT TAPIN 97890

Corps 24.

5768 LE PARIS-SPORT 1234

Corps 36.

23 VINS NATURELS 78

ÉGYPTIENNES DENTELÉES

Corps 14.

Ouverture SAMEDI 12 Juin

Corps 24.

**Le feu et l'eau sont
LIEU MALSAIN**

Corps 36.

**Le Moderne
ROCHELLE**

Corps 48.

**Courses
MORAL**

FONDERIE WARNERY

4, 6, 8, Rue Humboldt — PARIS

ITALIENNES INITIALES

Corps 12.

12345 LE PRÉSIDENT CARNOT, MORT 56789

Corps 16.

234 L'AFRIQUE CENTRALE VUE 567

Corps 20.

12345 UN CHORÉGRAPHE A 67890

Corps 28.

24 FUNICULAIRES 89

Corps 36.

23 TAPIS VERT 60

FANTAISIES DIVERSES

Corps 12, Nº 1

8123 LES COUPONS ÉMIS 4567

Corps 18, Nº 2

324 POLE NORD 561

Corps 24, Nº 3

A PASTEUR 458

Corps 36, Nº 4

509 - CHINE

Corps 12, Nº 198

COURS DE DANSE

Corps 14, Nº 206

128 LE SPITSBERG 504

Corps 20

26 ASSAUT 35

Corps 24

8 UN BAR 9

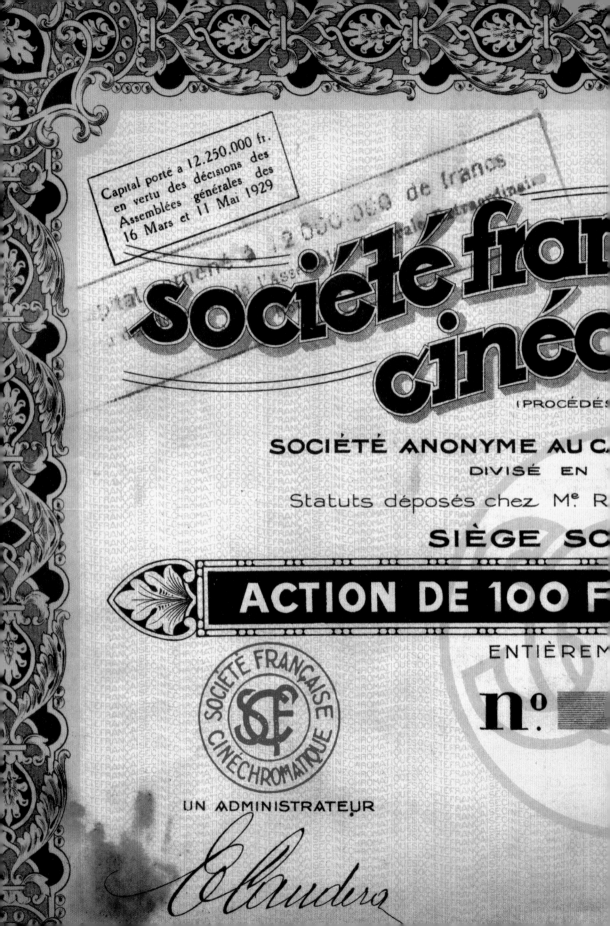

çaise

rromatique

(BERTHON)

...TAL DE 12.000.000 DE F^{cs}.

...00 ACTIONS

...t LETULLE Notaire à Paris

...AL A PARIS

...ANCS AU PORTEUR

LIBÉRÉE

331

Paris, le 1er Janvier 1929

UN ADMINISTRATEUR

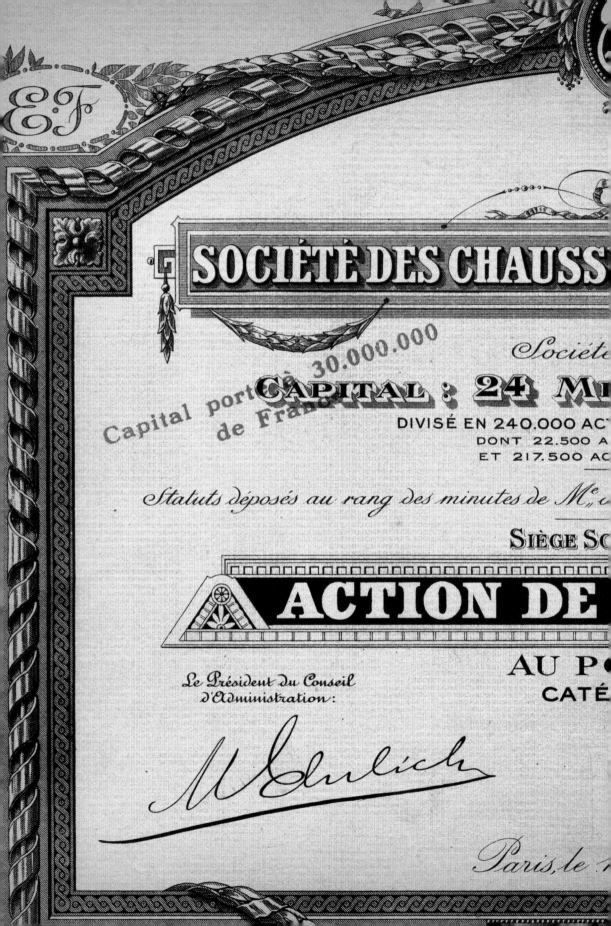

EF

SOCIÉTÉ DES CHAUSS

Sociét

Capital porté à 30.000.000
de Franc

CAPITAL : 24 M

DIVISÉ EN 240.000 ACT
DONT 22.500 A
ET 217.500 AC

Statuts déposés au rang des minutes de M^e

SIÈGE SO

ACTION DE

AU P
CATÉ

Le Président du Conseil
d'Administration :

M. Ehrlich

Paris, le

par décision de l'Assemblée Générale

Extraordinaire du 22 Août 1927

Porté à 65 millions de Francs

RES EHRLICH FRÈRES

nonyme

IONS DE FRANCS

DE 100 FRANCS CHACUNE

CATÉGORIE A.

CATÉGORIE B.

eaux, Notaire à Paris le 20 Décembre 1923

A PARIS

ENT FRANCS

TEUR

RIE B

Le Délégué Spécial du
Conseil d'Administration :

Février 1928

Droit de Timbre
acquitté par abonnement.
Avis d'autorisation inséré
au Journal Officiel
du 21 Janvier 1928

E.F

SOCIÉTÉ DES TEXT[...]

Société Anonyme au Capit[...]

Divisé en 400.000 Ac[...]

SIÈGE SOC[...]

Statuts déposés en l'Étude de Mᵉ AUB[...]

modifiés par les Assemblées Générales [...]

SOCIÉTÉ DES TEXTILES CHIMIQUES DU NORD ET DE L'EST

Action de Deux Cent Cin[...]

Un Administrateur,

Nᵒ

P. FORVEILLE IMPRI[...]

ES **CHIMIQUES** DU **N**ORD ET DE L'**E**ST

e **Cent Millions de Francs**

0 Francs chacune

. A PARIS

Notaire à Paris, le 9 Novembre 1927

dinaires des 9 Mai et 12 Juillet 1928

nte Francs au Porteur

.640

Un Administrateur
ou par délégation du Conseil d'Administration,

R.N_____

VIAN

LE MEILL
PETIT
LEFÈV

LU
PETIT-BEURRE
NANTES

GRAND PRIX :: PARIS 1900

DOX
LU

UR BISCUIT

BEURRE

E - UTILE

CH. LORILLEUX

et Cie

16 RUE SUGER · PARIS-6ᵉ

ENCRES

D'IMPRIMERIE

BLANC

A
PARTIR DU 3 JANVIER

ART

et Industrie

III

LE NUMÉRO : 150 FRANCS

A. TOLMER

M I S E

THEORY
PRACTICE
PAST...
PRESENT
ART...
DRAUGHTSMAN
PAINTING
ARCHITECTURE
DECORATION
GEOMETRY
IMAGES
THE BOOK
PUBLICITY
PRECEDENT
WRITING
PRINTING
TYPOGRAPHY
SLOGANS
PHOTOS
BLACK AND WHITE
FORM
ESPRIT
STYLE
SIMPLICITY
DARING
BALANCE
AVANT-GARDE
MASS

the theory and practice of lay-out

EN PAGE

THE STUDIO LTD.

HIER

AUJOURD'HUI

le blaireau

savon à barbe

et

la peau irritée....

LA
CREME
A RASER
RAZELYS

RASE ET LISSE

ÉESSE — PARIS.

LES INITIALES ÉGYPTIENNES CARRÉES

Les petits œils existent comme Capitales corps 5, 6 et 7

872 — Corps 6 — 12 fr.

LES DYNAMITEUSES DE L'AIR

890 — Corps 7 — 10 fr.

ARRONDISSEMENT D'AGEN

873 — Corps 8 — *8 fr. 50

LES NORMANDS

874 — Corps 10 — *7 fr. 50

QUIMPERLÉ

875 — Corps 12 — 7 fr.

LES CROISADES

876 — Corps 14 — 6 fr. 50

FRÉDEGONDE

877 — Corps 16 — 6 fr.   20 A : env. 5 k. 75

RÉVOLUTION FRANÇAISE

891 — Corps 18 — 5 fr. 50   20 A : env. 6 k. 15

EXPOSITION DE TABLEAUX

878 — Corps 20 — 5 fr.   12 A : env. 4 k. 60

HISTOIRE ROMANE

879 — Corps 28 — 4 fr. 50   10 A : env. 8 k.

DE PARIS A ROME

880 — Corps 40 — 4 fr.   10 A : env. 17 k. 50

IMPOSITION

881 — Corps 48 — *3 fr.   10 A : env. 24 k. 40

BOULOGNE

1084 — Corps 60 — 3 fr.   5 A : env. 20 k.

L'ÉCHO

205

NORD MAGAZINE

N

**MARS
1931
N° 39**
4ᵉ Année

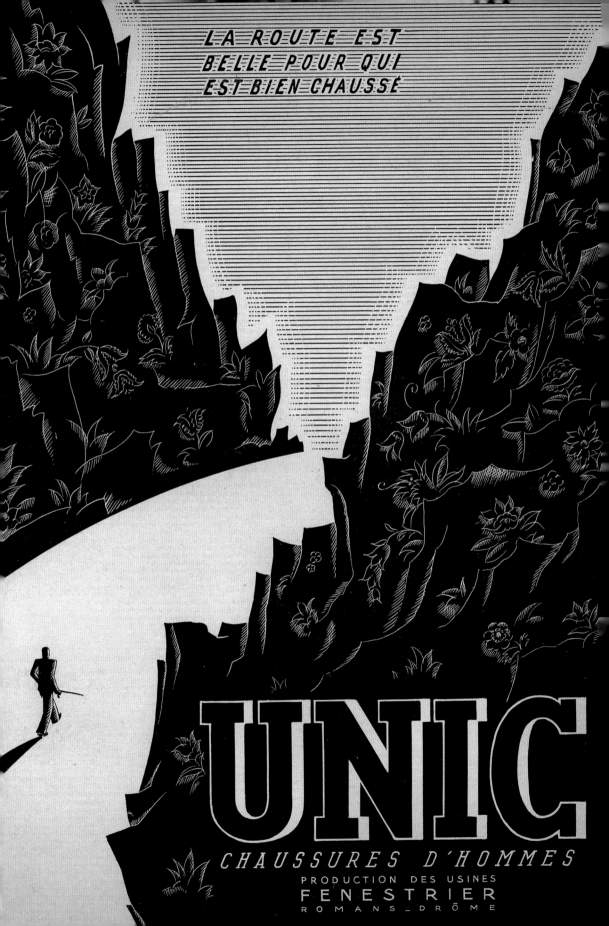

RENAULT 1939

4 CYLINDRES

LA JUVAQUATRE 4 PLACES
La voiture de votre travail et des loisirs en famille

7 litres
aux 100 kms
•
100 à l'heure

LA NOVAQUATRE 5 PLACES
La Puissance sans la dépense

10 litres
aux 100 kms
•
105 à l'heure

LA PRIMAQUATRE 5 PLACES
La voiture des joies sportives

11 litres
aux 100 kms
•
125 à l'heure
Conduite
intérieure
Coach
Cabriolet
Coupé

LA VIVAQUATRE 6 & 8 PLACES
La spacieuse voiture des familles nombreuses

12 litres
aux 100 kms
•
110 à l'heure

6 CYLINDRES

LA VIVA GRAND SPORT 6 PLACES
La Six cylindres imbattable en vitesse

15 litres
aux 100 kms
•
135 à l'heure
Conduite
intérieure
Coach
Cabriolet
Coupé

LA VIVASTELLA 8 PLACES
La plus belle et la plus confortable des 6 Cylindres

16 litres
aux 100 kms
•
130 à l'heure

8 CYLINDRES

LA SUPRASTELLA 6 & 8 PLACES
La plus somptueuse voiture

140 à l'heure

RENAULT   L'AUTOMOBILE   DE   FR

Confiance en l'avenir...

...YEZ CONFIANCE EN L'AVENIR... SURTOUT SI VOUS AVEZ CHOISI POUR VOITURE CELLE ...JI ASSURE A VOS ACTIVITÉS LE MAXIMUM DE RAPIDITE, D'ÉCONOMIE, DE SÉCURITÉ... UNE

RENAULT

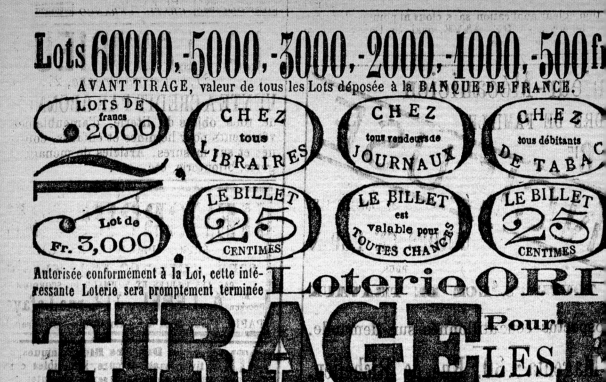

211

DIVERTISSEMENTS

5

PHARAON

DEBERNY
PEIGNOT

TYPOGRAPHIQUES

PHARAON GRAS

3 A	2414 - c. 60 init.	7 k. 70

MAIL

5 A	2397 - c. 48 init.	9 k. 20

SORTE
FLANS

5 A	2368 - c 48	6 k. 15

BRULER
12 34 56

8 A	2367 - c. 36	5 k. 00

FRANCHES
MILITAIRE

8 A	2366 - c. 30	3 k. 40

LUMIÈRES
ANTILOPE

30 A	2356 - c. 5	0 k. 70

QUELQUES BONS CONSEILS A CEUX QUI SE CONSACRENT A LA MODE FÉMININE

30 A	2357 - c. 6	0 k. 70

LES ATELIERS DE GRAVURE ET DE DESSIN D'UN PUBLICISTE MODERNE

30 A	2359 - c. 8	1 k. 10

CHOIX DE LIVRES D'ÉTUDE ET DE ROMANS RÉALISTES

25 A	2361 - c. 10	1 k. 10

IMPRESSION EN TAILLE-DOUCE, LITHOGRAPHIE

15 A	2362 - c. 12	1 k. 10

ÉDITION COMPLÈTE EN TROIS VOLUMES

15 A	2363 - c. 14	1 k. 50

FÊTE ET BAL DES JEUNES AUTEURS

15 A	2358 - c. 16	2 k. 00

SOLIDE AMITIÉ UN CAMARADE

10 A	2364 - c. 18	1 k. 80

FRAMBOISES LE PRESSOIR

10 A	2365 - c. 24	3 k. 10

BRÉVIAIRES
CONJUGUER

DEBERNY ET PEIGNOT PARIS

H

ERMÈS

28 Fg. St-Honoré
PARIS

MAILLOTS
DE BAIN
COSTUMES
DE PLAGE

Deuxième Année. — N° 21.   Prix : 30 centimes   15 Mai 1921.

LA BONNE SURPRISE

JOURNAL POUR TOUS LES ENFANTS

Paraît le 1er et le 15 de chaque mois.

Direction. Rédaction, Administration : 6, avenue Portalis, PARIS (8e).

ABONNEMENTS :

	FRANCE ET BELGIQUE	ÉTRANGER
Un an	6 50	8 francs.
Six mois	3 50	4 50

Pour la Belgique : Abonnements et correspondance, s'adresser :
14, rue de la Chancellerie, BRUXELLES.

JEAN-PIERRE IMPRIMEUR

ARTS

ET MÉTIERS

★

GRAPHIQUES

★

★

50

ARMAGNAC

BARNABÉ

CONDOM GERS

ADAM

LA REVUE DE L'HOMME

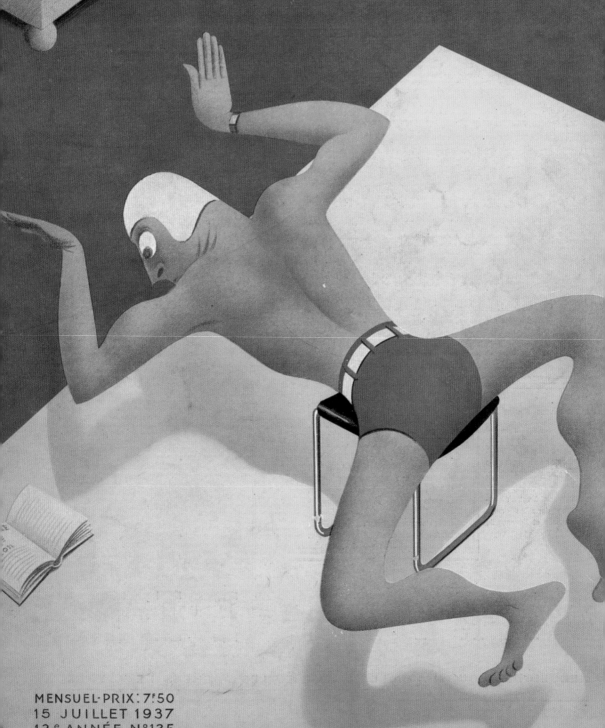

MENSUEL·PRIX: 7f50
15 JUILLET 1937
13e ANNÉE No135

223

Spécial Pays-Bas

LIBRA STUDIO

abcd

klmno

t y

efghij

ppqrs

vxyz

drolm

171. Travel poster, Renault, 1936.

172–73. Specimen pages of Égyptiennes Carrées, *Fonderie Typographique Mayeur*, 1905.

174. Plate of Lettres Bronze, *Album de Lettres*, N. Glaise, 1884.

175. Specimen page of Les Italiennes Allongées, *Fonderie Typographique Mayeur*, 1905.

176–77. Periodical mastheads for *Le Rappel*, March 24, 1909; *Le Bon Vivant*, March 30, 1901; *Le Monde Pittoresque*, October 5, 1884.

178. Periodical cover for *Le Petit Journal*, May 18, 1906.

179. Periodical cover for *Le Monde Comique*, date unknown.

180. Periodical masthead for *Le Père Gérard*, October 26, 1879. Book cover for *Drôleries Contemporaines*, date unknown.

181. Periodical cover for *Gil Blas*, illustration by Théophile Alexandre Steinlen, December 11, 1896.

182. Periodical cover for *La Nouvelle Revue Parisienne*, April 2, 1898.

183. Catalog, Société Anonyme de Construction de la Villette, 1880.

184–85. Product labels, *c.* 1900.

186–87. Specimen pages of Egyptiennes Allongées and Egyptiennes Dentelées, *Fonderie Warnery & Cie: catalogue général Warnery*, 1934.

188. Specimen page of Italiennes Initiales, *Fonderie Warnery & Cie: catalogue général Warnery*, 1934.

189. Specimen page of Fantaisies Diverses, *Fonderie Warnery & Cie: catalogue général Warnery*, 1934.

190–91. Decorative alphabet, date unknown.

192–93. Stock certificate, Société Française Cinéchromatique, January 1929.

194–95. Stock certificate, Société des Chaussures Ehrlich Frères, February 1928.

196–97. Stock certificate, Société des Textiles Chimiques du Nord et de l'Est, November 1927.

198–99. Metal sign, Viandox, *c.* 1935. Cardboard sign, LU Petit-Beurre, 1930s.

200. Advertisement for Ch. Lorilleux & Cie. printing firm, *Arts et Métiers Graphiques* 48, 1935.

201. Catalog cover for "Blanc" at Les Trois Quartiers, designed by Marian Andrew, date unknown.

202. Periodical cover for *Art et Industrie*, April 1946.

203. Book cover for the English edition of *Mise en Page*, 1931.

204. Point-of-purchase advertisement for Razelys, *c.* 1930.

205. Specimen page of Les Initiales Égyptiennes Carrées, *Fonderie Typographique Mayeur*, 1905.

206. Magazine cover for *Nord Magazine* No. 39, designed in 1927 by A.M. Cassandre, published March 1931.

207. Advertisement for UNIC, 1932.

208–9. Posters, Renault, 1939.

210–11. Advertisements, 1920s.

212–13. Specimen pages of Pharaon, Deberny et Peignot, *Divertissements Typographiques* #5, 1935.

214–15. Advertisement for Hermès, Deberny et Peignot, *Divertissements Typographiques* #5, 1935.

216. Catalog cover for dress store Marlène, *c.* 1925.

217. Periodical cover for *La Bonne Surprise*, May 15, 1921. Box top for children's printing press, Jean-Pierre Imprimeur, *c.* 1925.

218–19. Initials from *La Broderie Lyonnaise*, April 1, 1950.

220. Magazine cover for *Arts et Métiers Graphiques* 50, 1935.

221. Advertising sign for Armagnac Barnabé, *c.* 1935.

222. Magazine cover for *Adam: La Revue de l'Homme*, designed by Paolo Garetto, July 15, 1937.

223. Magazine cover for *Publimondial* No. 9, designed by Ternat, 1947.

224–25. Magazine covers for *Publimondial* Nos. 31 and 35, designed by Libra Studio and Lucien Lorelle, 1948.

226–27. Alphabet from *La Lettre Artistique et Moderne*, designed by Draim, *c.* 1930.

DUTCH

THE TRUTH ABOUT DUTCH SLAB SERIF TYPEFACES IS THAT THEY were not exactly Dutch. Most slab faces in Holland were taken directly from or influenced by a surge of existing designs from elsewhere in Europe (particularly from Germany's D. Stempel foundry), with minor changes in wood and metal recreations. At first, square serif types were sporadically employed by Dutch commercial job printers and were primarily used for small headlines, to add boldness to a printed page. Their use on product labels (pp. 242–45) shows how readable they are. In the twenties and thirties the leading commercial arts trade journals (including *De Reclame*) presented their subscribers with a number of eclectic ways of customizing square serif letters to make them timely and stylish. During this period, Dutch lettering books were rich with quirky and elegant customizations for posters, packages, and editorial designs.

Slabs were not as fashionable as Holland's famous indigenous typography, which developed through its modern wing, the Amsterdam School, and was promoted through the magazine *Wendingen*. This style was a cacophonous mix of different elements—serif and sans serif, linear and curvilinear—and was described by critics as "artistic, decorative, symbolic, fantastic, antisocial, lyrical, passive, romantic, aesthetic, craftsmanly." Evolving in opposition to this style was the increasingly popular New Typography movement. Inspired by Bauhaus experiments and the avant-garde approach of de Stijl, its adherents advocated functional and practical sans serif types. However, despite these rivals, slab serifs such as Egyptian Bold Condensed (right) showed they still had modern applications in addition to lending a page weight and colour. Modern designers in the twenties and thirties realized that serifs increased legibility by tying letters together and making word units easier to recognize. These geometric forms enabled what Douglas C. McMurtrie called effective comprehension for "hurried readers in a fast-moving age." In Holland's growing advertising field of the early twenties, designers had been forced to make do with old-fashioned gothics. As McMurtrie wrote: "There was obviously a crying need for types designed on modern principles. Enter the slab serif to temporarily fill the gap."

11e JAARGANG No 8
23 FEBRUARI 1932

Reclame

verschijnt elke week

REDACTIE EN ADMINISTRATIE

HOFWIJCKSTRAAT 9 DEN HAAG

No. 2749 Corps 72/60   7 A, 15 a   Minimum 24 Kg.

Redacteur
DE KUNST

No. 2748 Corps 60/48   7 A, 15 a   Minimum 20 Kg.

Cours Public
ANNONCES

No. 2747 Corps 48   7 A, 15 a   Minimum 16 Kg.

Kunstnijverheid
STEMPELDRUK!

No. 2746 Corps 36   11 A, 25 a   Minimum 14 Kg.

Archiv für Heraldik
BUCH DER JUGEND

N.V. LETTERGIETERIJ „AMSTERDAM" VOORHEEN N. TETTERODE

1866

GHB

HOLLAND

PAPIERMOLEN

DE RECLAME

Wanneer een drukker zijn vak verstaat en over een goede installatie beschikt, dan kan hij met weinig middelen — met weinig kleuren — een prachtig, opvallend stuk werk voor U maken. Ziehier een klein voorbeeld van hetgeen wij voor U maken kunnen. Vlug, goed en efficient — zooals U dat van een groote, modern-geoutilleerde offset-drukkerij kunt verwachten

BOEK- EN STEENDRUKKERIJ Vⁿ A. FLACH - SNEEK

Alle aanvragen te richten aan den Algem. Vertegenwoordiger: C. F. HUIJER, Wouwermanstraat 19, Amsterdam. Telef. 22121

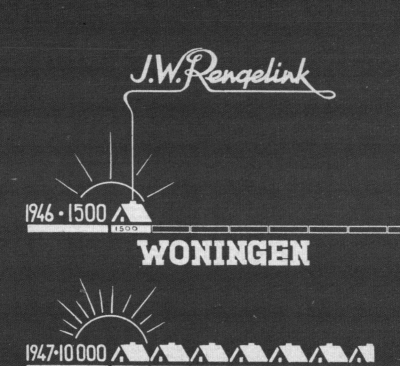

WONINGEN

UITGAVE: PARTIJ VAN DE ARBEID · TESSELSCHADESTRAAT 31 · AMSTERDAM-W

design and PRINTING IN THE NETHERLANDS

L. P. Hoogendijk

Gouda

Hoogstraat.

VROOM & DREESMANN

2 ROLLETJES PRIMA BOORLINT

NEDERLAND

HET GEREGELD GEBRUIK VAN KOEK
BEVORDERT
DE GEZONDHEID
12 1/2
cts
C. JAMIN
ROTTERDAM.

IK EET H-O
IK NIET

SOLVAY

NAT. HANDEL- INDUSTRIE- EN NIJVERHEIDSTENTOONSTELLING

NAHINTO

R.A.I.
Gebouw
AMSTERDAM 26 Aug. t/m 6 Sept.

SOLVAY & C.ie

K 2193

Solvay

N.V. NEDERLANDSCHE
ROTOGRAVURE

MAATSCHAPP
LEIDE
GALGEWATER 2

231. Specimen page of Egyptian, Amsterdam Continental Types and Graphic Equipment, 1950s.

232–33. Advertisements from the *Amsterdamsche Courant*, December 1882.

234. Magazine cover for *De Reclame*, February 1932.

235. Specimen page of Vette Atlas, N.V. Lettergieterij Amsterdam, 1919.

236. Promotional brochure for GHB paper mill, July 1935.

237. Magazine cover for *De Reclame*, July 1931.

238–39. Advertisement for Flach, 1931.

240. Brochure cover, designed by André, 1947.

241. Book cover for *Design and Printing in the Netherlands*, date unknown.

242–45. Product labels, 1900–1952.

246. Advertisement for N.V. Nederlandsche Rotogravure, *De Reclame*, July 1931.

BRITISH

Britain was a wellspring of slab serifs and other titling typefaces. During the early nineteenth century leading British type-founders, including Robert Thorne and William Thorowgood, were known for their indelicate bastardizations of familiar faces, fattening, elongating, outlining, and shadowing classic typefaces such as Bodoni and Garamond for advertising display purposes. In 1845 Clarendon, designed by Robert Besley for Thorowgood and Co., was produced and registered under Britain's Ornamental Design Act. This legal protection expired after three years and Clarendon became one of the most commonly copied slab serif types in the country. There was arguably a larger number of slab serif typefaces on a wider range of printed ephemera emanating from the British Isles than from any other industrial country. The British also had their own terminology: Although the rest of the typographic world referred to the majority of slab faces as "Egyptian," in Britain "Antique" was the designated term. Antiques eventually came in dozens of widths and heights that routinely filled lines with a cacophony of size variations (pp. 258–63).

The first English slab serif printing type was commercially introduced in 1815 by Vincent Figgins, whose foundry was also famous for the creative alchemy of turning engravings on wooden logs into legible letterforms. Among the general public, the exaggerated fat face and bold slab type phenomenon was seen as an outgrowth of the mass-produced ugliness fostered by the Industrial Revolution. However, some vocal type users heralded it as a brilliant typographical innovation. Another strong advocate for the slab aesthetic was Stephenson, Blake & Co. in Sheffield, England, which opened its doors in 1818 and issued alluring specimens over the course of many decades. The foundry stocked numerous weights of French Antique and was responsible for reintroducing the Egyptian elongated and expanded faces that became popular for editorial headlines. It also created its own version of Clarendon, called Consort (pp. 274–75). But it was *Typography*, the quarterly journal of the Shenval Press, that most loudly sounded the trumpet for the sculptural beauty and typographic clarity of slabs, using them as the building blocks for many of its covers throughout the late 1930s.

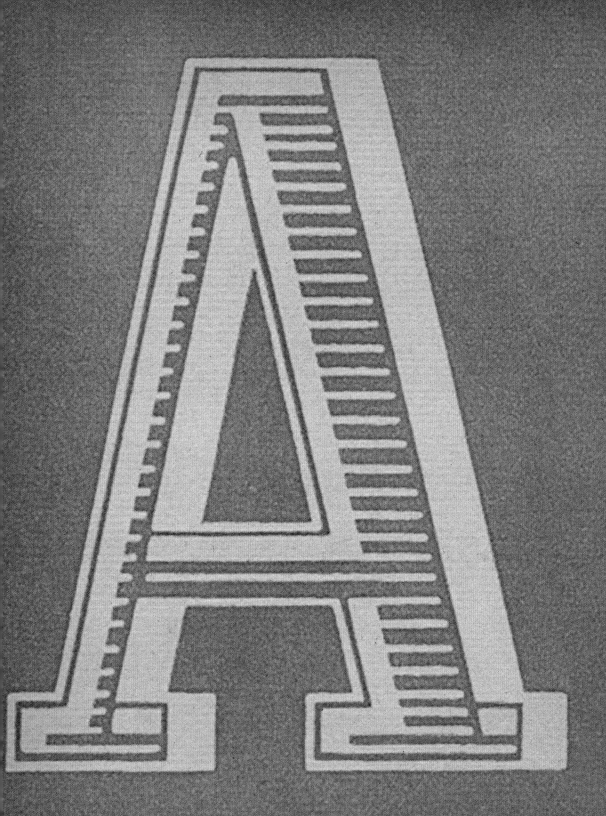

ART AND TECHNICS

HARVEST { THE STORY OF MY LIFE, BY FATHER GAPON.
NUMBER { "LA FAYETTE" BY MAX PEMBERTON, ETC. ETC.

THE STRAND MAGAZINE

PRICE   10 CENTS

Nº 176 VOL 30

SEPT. 1905

NEW YORK: THE INTERNATIONAL NEWS COMPANY.   LONDON: GEORGE NEWNES LIMITED.
Toronto: THE TORONTO NEWS COMPANY.   Montreal: THE MONTREAL NEWS COMPANY

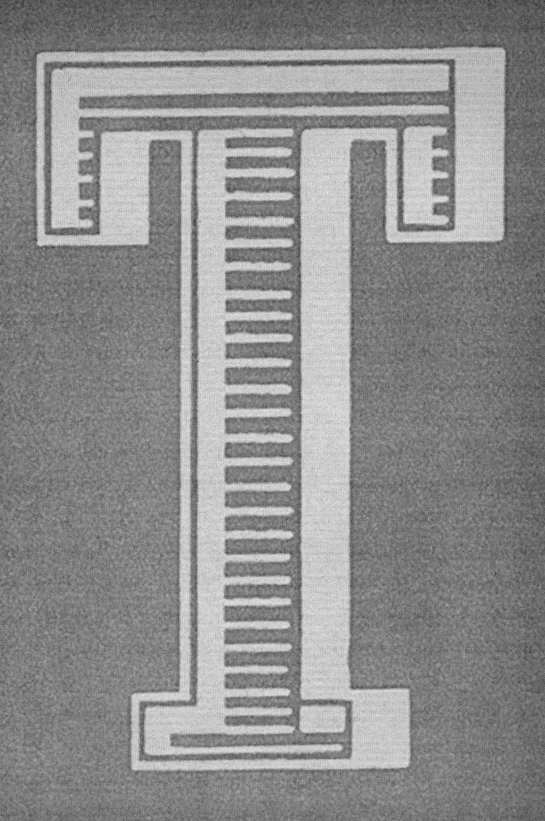

AUTUMN 1950

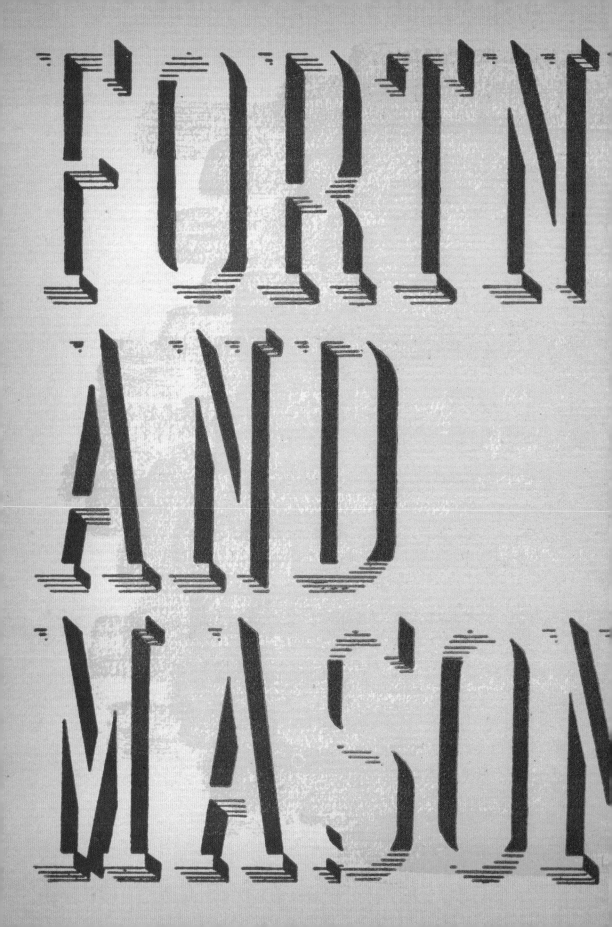

TIME

SPRING 1933

COLLECTION

THE

BIRKE

Lessees and Managers ... ...

MONDAY, S

AND DURIN

Starring Engagement and First Appeara

SIS

PHIL

AND BRO

DIEN

TRE

HEAD.

... ...   Messrs. CLARKE & RILEY

PT. 3, 1888

THE WEEK.

in Birkenhead of the only and Original

ERS

LIPS

HERS

5

TYPOGRAPHY

Typography: a quarterly published by the Shenval Press

Two Shillings: Spring 1938

TYPOGRAPHY: SUMMER 1939: TWO SHILLINGS

TYPOGRAPHY 8

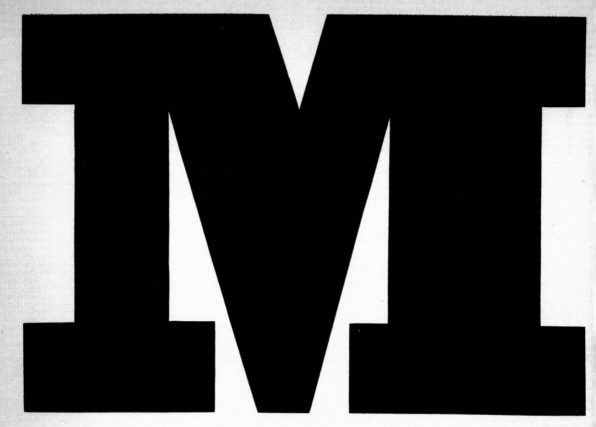

TWENTY LINES ANTIQUE, No. 1

OTHER SIZES SUPPLIED WITH EQUAL FACILITY

STEPHENSON, BLAKE & CO. LIMITED

SHEFFIELD, LONDON, MANCHESTER

HORN

TWELVE LINES ELONGATED ANTIQUE, No. 2

MERCHANT

EIGHT LINES ELONGATED ANTIQUE, No. 2

NORTH STAND 15

OTHER SIZES SUPPLIED WITH EQUAL FACILITY

STEPHENSON, BLAKE & CO. LIMITED

SHEFFIELD, LONDON, MANCHESTER

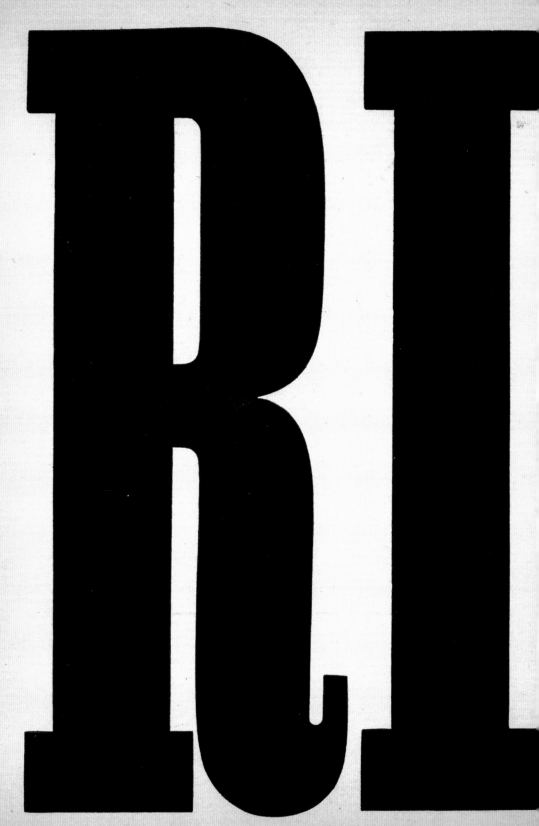

HARM

CHARGES

OTHER SIZES SUPPLIED WITH EQUAL FACILITY

STEPHENSON, BLAKE & CO. LIMITED   SHEFFIELD, LONDON, MANCHESTER

OCEANIC

GREYHOUND
Marine Race

SEA MERCHANT
Bond Number 20

OTHER SIZES SUPPLIED WITH EQUAL FACILITY

STEPHENSON, BLAKE & CO. LIMITED · SHEFFIELD, LONDON, MANCHESTER

Stephenson Bla

e · **Sheffield**

WHO'S WHO IN ART

WHO'S WHO IN ART

6TH EDITION

ART TRADE PRESS

WHO'S WHO IN ART

Biographies of eminent artists,
designers, craftsmen, critics, writers,
teachers, collectors, curators.
With an appendix of signatures.

SIXTH EDITION

DESIGNERS

CRAFTSMEN

CONTRACTORS

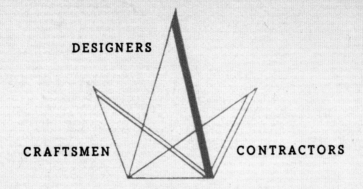

COCKADE  LIMITED

DISPLAYS

EXHIBITIONS

RECONSTRUCTION

MODELS

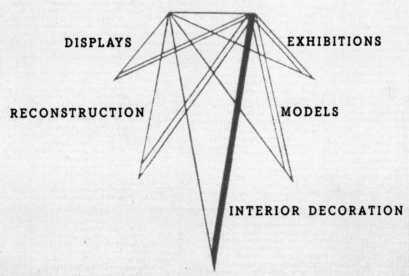

INTERIOR DECORATION

35 THURLOE PLACE LONDON SW7 KEN 4824

EGYP

EXPA

OPEN

NDED

TIAN

TIAN
NDED
AND

STEPHENSON
BLAKE
SHEFFIELD

EXPA
EGYP

EGYPTIAN EXPANDED

A VICTORIAN REVIVAL

USEFUL IN ALL FORMS OF

BOLD

DISPLAY

TO WHICH HAS BEEN ADDED

THE NEWLY DESIGNED

EGYPTIAN EXPANDED

OPEN

IN THREE SIZES

•

SB

STEPHENSON BLAKE

THE CASLON LETTER FOUNDRY

SHEFFIELD 3 · ENGLAND

& 33 ALDERSGATE STREET EC1

EGYPTIAN EXPANDED AND EGYPTIAN EXPANDED OPEN

··

Sole Agents

AMERICAN WOOD TYPE MFG. CO.

42-25 NINTH STREET
LONG ISLAND CITY-1 N.Y.

RAvenswood 9-5779

Cable Address: Woodtype, New York

EGYPTIAN EXPANDED

36 Point

ABCD
EFGH
abcdef

30 Point

IJKLM
NOPRS
ghijkm
nopqrs

36 Point

ABCDEFGH

30 Point

QRSTUVWX

24 Point

ABCDEFGHIJKI
abcdefghijklmnop

TUVWX
tuvwxyz
2345678

ABCDEFGH
IJKLMNOP
abcdefghijkl

QRSTUVWXY
ZÆŒ&23456£$
abcdefghijklm
nopqrstuvwxz

IJKLMNOP
YZ-123456789
MNOPQRSTUVX
rstuvwxyz&ﬁﬂﬀﬄﬃ

36 Point

ABCDEFGHJK

abcdefghijklmno

pqrstuvwxyzæœ

30 Point

ABCDEFGHIKLM

NOPQRSTUVWX?

abcdefghijklmnopqr

12 Point

ABCDEFGHIJKLMNOPQRSTUVWXYZ
ÆŒ&1234567890£$,.:;-'!?ABCDEFGHIJ
abcdefghijklmnopqrstuvwxyzæœfifffflffiffl

10 Point

ABCDEFGHIJKLMNOPQRSTUVWXYZÆŒ,.:;-'!?
abcdefghijklmnopqrstuvwxyzæœfifffflffiffl&1234567
With his breadth of vision and high conception of
civic duties, Henry Stephenson soon stood out in
the Council Chamber, and his fellow townsmen
were not slow to appreciate the abilities of their

CONSORT BOLD

ABCDEFGHIJ
abcdefghijklm

KLMNOPQRSTU
nopqrstuvwxyz?

CONSORT CONDENSED

ABCDEFGHIJKLMN
abcdefghijklmnopqrst

OPQRSTUVWXYZÆŒ,.:;!?
tuvwxyzæœ&fiffflffiffll1234

He decided to make the army his career. He took
his first commission as Cornet in the 2nd Dragoons
(the Scots Greys) in July 1857. It was in March

60 Point 5 A, 8 a ; about 33 lb.

BRIGHTON

Deposit Banks

48 Point 7 A, 12 a ; about 25 lb.

MASON
merchants

36 Point 9 A, 16 a ; about 22 lb.

BANKING
Many branch

30 Point 12 A, 24 a ; about 20 lb.

ERITH BANK
Numerous branch
offices were being

24 Point 20 A, 44 a ; about 18 lb.

DEPOSIT BANKS
The aggregate value of
guaranteed investment

18 Point 26 A, 54 a ; about 15 lb.

DEPARTMENTS
for the convenience of
patrons changing their

12 Point 44 A, 90 a ; about 12 lb.

NATIONAL SOCIETIES
with foreign branches present
great facilities to travel parties

10 Point 56 A, 114 a ; about 10 lb.

MUNICIPAL CORPORATION
Having many Branch Establishments
throughout Ireland, and more than 230
Correspondents in Britain and abroad

8 Point 66 A, 130 a ; about 8 lb.

THE NATIONAL AND PROVINCIAL
having numerous Branch Establishments in
England and 360 correspondents at home and
abroad gives facilities to those about to travel
or are contemplating a change of residence

6 Point 70 A, 150 a ; about 5 lb.

NATIONAL AND PROVINCIAL BANKS
having numerous Branch Establishments throughout the
country, and 380 Correspondents at home and abroad,
offer special facilities to parties changing their residence or
about to travel, as well as to officers of the Army and Navy

5 Point 54 A, 110 a ; about 4 lb.

METROPOLITAN AND PROVINCIAL BANKS
having numerous Branch Establishments throughout England
and over 290 Correspondents at home and abroad, offer special
facilities to parties changing their residence or about to travel,
as well as to officers of the Army and Navy for receiving or sending
their remittances, and last year the aggregate value of guaranteed

72 Point   5 A, 6 a   about 32 lb.

MERCHANT
British Printers

18 Point   32 A, 66 a ; about 15 lb.

PROVINCIAL BANK
has Branch Establishments
with Correspondents 26850

12 Point   56 A, 112 a ; about 12 lb.

METROPOLITAN BANKERS
prosperous Branches throughout the
kingdom, with 28460 Correspondents
at home and abroad, give facilities to

10 Point   64 A, 140 a ; about 10 lb.

THE BANKING CORPORATION
having a prosperous Branch Establishment
n Amsterdam, and over 2950 Agents in the
Colonies, gives great facilities to patrons at
home and abroad who intend changing their

8 Point   72 A, 150 a ; about 8 lb.

METROPOLITAN AND PROVINCIAL BANK
having numerous Branch Establishments in England
gives greater facilities to parties changing residence,
or intending to travel, as well as for the receiving or
forwarding of Army and Navy remittances, and last
year the aggregate value of guaranteed deposits were

6 Point   80 A, 170 a ; about 5 lb.

EDINBURGH AND PROVINCIAL CORPORATIONS
numerous Branches established throughout England, and 2830
correspondents at home and abroad, this Bank gives facilities to
parties about to change their residence or intending to travel, as
well as to Officers of the Army and Navy for receiving and sending
remittances, and for last year the aggregate value of guaranteed
deposits amounted to £25,470, showing a decided increase over

48 Point   9 A, 16 a ; about 26 lb.

BRUSHES
Finest Camel

36 Point   10 A, 20 a ; about 21 lb.

PROVINCIAL
Deposit in Banks

30 Point   16 A, 34 a ; about 20 lb.

NATIONAL BANK
Branch establishments
in over 260 towns and

24 Point   26 A, 50 a ; about 18 lb.

BIRMINGHAM BANKS
have established Branches in
numerous important towns for

ABCDEE

MNOPQR

Zabcdefgh

GHIJKL
TUVWXY
jklmnopq

LETTER FOUNDRY · SHEFFIELD

EASTERN RESIDENCE
Furnished Complete
Select Bowling Green

MODERN MUSICAL RECORDS
English Dramatic Company
Full Chorus of superb Artistes

SPECIMENS FROM BRITISH CONTRIBUTORS
Instructive and Amusing Volume
Numerous Performances in December

PROMINENT CAPITALS

Point Line enables a Compositor to set Antique and similar faces with an ordinary Modern or Old Style, ensuring correct alignment

OFFICES FURNISHED COMPLETE

Point Line enables the Printer to compose Antique and similar faces with ordinary Old Style and Modern, and ensures the perfect alignment without resorting to those crude methods employed under the old standard. Ask for Point Line Chart showing 2345678

DESIRABLE ANTIQUE SERIES

Point Line enables the Printer to set up Antiques and similar faces with ordinary Modern or Old Style, and ensures perfect alignment without resorting to card, etc.

STEPHENSON, BLAKE & Co. Ltd. SHEFFIELD

Point Line enables the Printer to set up Antiques and similar faces with ordinary Modern or Old Style, and ensures perfect alignment without resorting to the crude methods employed under the old dispensation. Point Line Charts, showing exact amount of packing required to line one body with another will be sent post free upon receipt

MONTAGU

MONTAG
LONDO

ASSOCIATED

ME

TWENTY LINES FRENCH ANTIQUE, No. 2

IRON

249. Magazine cover for *Art and Technics*, Autumn 1950.

250. Magazine cover for *The Strand Magazine*, September 1905.

251. Interior page from *Art and Technics*, Autumn 1950.

252–53. Catalog cover for Fortnum & Mason, designed by E. McKnight Kauffer, Spring 1933.

254–55. Detail of playbill, Theatre Birkenhead, September 3, 1888.

256–57. Magazine covers for *Typography* Nos. 5 and 8, 1938–1939.

258–63. Specimen pages for the Antique type family, Stephenson, Blake & Co., 1924.

264–67. Catalog pages, Stephenson, Blake & Co., *c.* 1947.

268–73. Specimen sheet of Egyptian Expanded and Egyptian Expanded Open, Stephenson, Blake & Co., 1960s.

274–75. Specimen pages of Consort, Consort Bold, and Consort Condensed, Stephenson, Blake & Co., 1924.

276–77. Specimen pages of Antique and Condensed Antique No. 3, Stephenson, Blake & Co., 1924.

278–79. Specimen of the Playbill range, Stephenson, Blake & Co., 1924.

280–81. Specimen of Antique, Stephenson, Blake & Co., 1924.

282–83. Luggage label, Montague Hotel, 1930s.

288. Specimen page of the Antique type family, Stephenson, Blake & Co., 1924.

GERMAN

S LAB SERIFS FOUND OPEN ARMS AND NIMBLE TYPESETTING fingers in Germany prior to the turn of the twentieth century. Then, after World War I, the subset of geometric or flat serifs became an essential part of the experimental Modern and quotidian commercial design toolbox. Type designers discovered ways to use serifs that were absurd (p. 296) and sometimes hilarious too (p. 298). These were not just uniform block serif styles but nuanced variations that could actually influence viewers' attitudes and make them read messages in a particular way. For example, the two *Gebrauchsgraphik* covers (pp. 302–3) show slabs as both warmly old-fashioned and coolly modern.

In Germany slabs symbolized technological progress. In 1930 the D. Stempel foundry published an advertisement in *Archiv für Buchgewerbe*, one of Germany's leading trade magazines, for Memphis Light and Bold (later called Girder when exported to the American market). The typeface was specified as appropriate for cosmetics and groceries (in light weight), and heavy industrial products, motor supplies, and household fixtures (in heavy weight). A later advert stated that Memphis types are related to Egyptians, but in form and proportion they are modern. Stempel developed more weights of Memphis for decidedly more functionality, spawning Memphis-like faces at other German foundries. Included in the renaissance of the slab were Welt-Antiqua at Ludwig & Mayer in Frankfurt am Main (1931), Beton, designed by Heinrich Jost for Bauersche Gießerei, also in Frankfurt, and City, designed by Georg Trump for H. Berthold AG in Berlin (both 1930). The Germans also popularized the modern genre internationally. After Girder was released, Chicago-based foundry Ludlow released Karnak and Obelisk. In 1931 the American Type Founders produced Stymie and by 1933 Memphis had been brought out by The Mergenthaler Company in Brooklyn. After World War II, many trade journals featured slabs on covers and inside layouts, aiming to show that even old styles could be made contemporary. The *Graphik* logo (p. 324), for instance, a souped-up Playbill, had its heavy slabs at the top and bottom of letters joined into a monoslab that smartly complements the collage illustration, while the *Die Mappe* logo (pp. 326–27) was customized to make it more streamlined.

Othello

3B 282 cedar Othello Swan Pencil Co GERMANY

Schwan-Bleistift-Fabrik · Nürnberg
Swan Pencil Co. · Nuremberg, Bavaria

ABCDE
FGHIJK
LMNOP
QRSTUV
WXYZ

Schriftgiesserei Jul. Klinkhardt, Leipzig.

ABCD
KLMNO
VW
abcdefg
pqrst

Kursiv.

ABCDEFGHI
JKLMNOPQRS
TUVWXYZ
abcdefghijklmn
opqrstuvwxyz
1234567890

Schriftgiesserei Schelter & Giesecke, Leipzig.

Schriftgiesserei I

ABCDEFG
PQRST
abcdefghijk
12345 vw

Schriftgiesserei I

tienne.

FGHIJ
QRSTU
YZ
ijklmno
wxyz

omp., Stuttgart.

ABCDEF
GHIJKLMN
OPQRSTUV
WXYZ

Schriftgiesserei Genzsch & Heyse, Hamburg.

Schattierte Jonisch.

ABCDEFGHI
JKLMNOPQRST
UVWXYZ
abcdefghijklmno
pqrstuvwxyz
1234567890

Schriftgiesserei Flinsch, Frankfurt a. M.

enne.

HIJKLMNO
VWXYZ
mnopqrstu
yz 67890

Comp., Stuttgart.

Verl. v. Jul. Hoffmann, Stuttgart.

M N O P Q R S
U V W X Y Z

riftgiesserei von Schelter & Giesecke in Leipzig.

K L M N
T U V
H A M

Schriftgiesserei von

D E F G H I J

4

N O P Q R S T

V W X Y Z

ISENKARTE.

von Julius Klinkhardt in Leipzig.

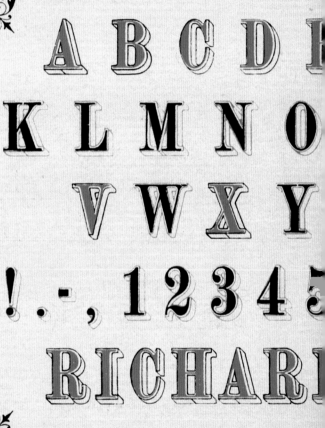

A B C D E
K L M N O
V W X Y
! . - , 1 2 3 4 5
RICHARD

Schriftgiesserei

5

D E F G H I J

N O P Q R S T

V W X Y Z

RCELONA

von Genzsch & Heyse in Hamburg.

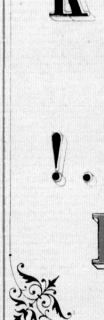

9

C D E F G H I J K L

O P Q R S T U V W

45 X Y Z 6 7 8 9 0

d e f g h i j k l m n

A B C D E
M N O P Q
1 2 3 4 5 X
a b c d e f

P Q R S
X Y Z &
URG

& Giesecke in Leipzig.

Schriftgiesserei von Schelter & Giesecke in Le

6

F G H I J
P Q R S T U
Ä Ö Ü &
7 8 9 0 ; : ' ?
WAGNER

in Frankfurt a. M.

A B C D E F
K L M N O P
U V W X
MEMORAN

Schriftgiesserei von Roos & Junge

A B C D E F G
L M N O P Q R
1 2 3 4 5 W X Y Z
a b c d e f g h i
o p q r s t u v

Schriftgiesserei von Genzsch & H

10

G H I J K L
S T U V W
Z 6 7 8 9 0

A B C D E F G H
L M N O P Q R S
V W X Y Z

ABCDEFGHIJKL
MNOPQRSTUVW
XYZ
123456
7890

Schriftgiesserei von Genzsch & Heyse in Hamburg.

Italienne Kursiv.

ABCDEFGH
IJKLMNOPQR
STUVWXYZ

Schriftgiesserei von Julius Klinkhardt in Leipzig.

ABCDE
MNOPQ
abcdefX
mnopqrs

Schriftgiesserei

ABCDEFG
OPQRSTU
abcdefghijk
12345 vw

Schriftgiesserei von

A B C D E F G H I J
K L M N O P Q R S T U
V W X Y Z

1 2 3 4 5 6
7 8 9 0

Schriftgiesserei von Genzsch & Heyse in Hamburg.

...ienne.

G H I J K L
S T U V W
Z ghijkl
u v w x y z

Frankfurt a. M.

Italienne Kursiv.

A B C D E F G H I J
K L M N O P Q R S T U
V W X Y Z
a b c d e f g h i j k l m n o p
q r s t u v w x y z
1 2 3 4 5 6 7 8 9 0

Schriftgiesserei von Schelter & Giesecke in Leipzig.

...nne.

H I J K L M N
V W X Y Z
m n o p q r s t u
y z 6 7 8 9 0

Giesecke in Leipzig.

Verl. v. Jul. Hoffmann, Stuttgart

ABCD
JKLM
RSTU
abcde
lmnopq
12345 xy

E F GH
NOPQ
VWXY
Z fghik
rstuvw
z 67890

ABCDE
LMNOP
VWXYZ
abcdefg
pqrstuv
verkauf 123

FGHJK
QRSTU
Pinselschrift
niklmno
vxyz Aus-
4567890

A B C D

K L M N

S T U V

a b c d e f

n o p q r s

1 2 3 4 5

EF GHI
OPQR
WXYZ

ghiklm
uvwxy
tz 67890

ABCDE
LMNO
UVW

abcdefg
nopqrs
y12345

F G H I K
P Q R S T
X Y Z

g h i j k l m
t u v w x
6 7 8 9 0 Z

GEBRAUCHSGRAPHIK

JULI
1937

International Advertising Art

Zapf 37

Herausgeber: Prof. H. K. Frenzel, Editor
Frenzel & Engelbrecher „Gebrauchsgraphik"
Verlag, Berlin SW 68, Wilhelmstraße 148.
Alleinvertreter für U. S. A. und Kanada:
The Book Service Company, 15 East 40th Street
New York City, Sole Representatives for the
United States of America and Canada

GEBRAUCHSGRAPHIK

NOVEMBER 1937

Herausgeber : Prof. H. K. Frenzel, Editor
Frenzel & Engelbrecher „Gebrauchsgraphik" Verlag, Berlin SW 68, Wilhelmstr. 148

Alleinvertreter für die Vereinigten Staaten von Amerika und Kanada:
The Book Service Company, 15 East
40th Street, New York City, U.S.A. Sole
Representatives for the United States
of America and Canada

INTERNATIONAL ADVERTISING ART

A B C D E F
G H I J K L M
N O P Q R S
T U V W X

abcdefghij
klmnopqrs
tuvwxyz
0123456789

A B C D

J K L M

R S T U

abc def

nopqrst

12345

E F G H

N O P Q

V W X Y

Z ghiklm

U V W X Y Z

67890

VERLAG FRANCKEN & LANG G.M.B.H. BERLIN W 30

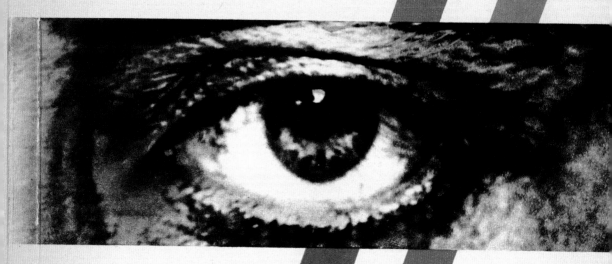

2. AUGUSTHEFT 1929 HEFT NR.

DIE REKLAME

ZEITSCHRIFT DES DEUTSCHEN REKLAME-VERBANDES E.
(VERBAND DEUTSCHER REKLAMEFACHLEUTE E.V.

HERAUSGEBER: PROFESSOR H·K·FRENZEL·EDITOR ✝

SEPT.

VERTRIEB FÜR U·S·A & CANADA: THE BOOK SERVICE COMP. NEW YORK ✝

GEBRAUCHS
·GRAPHIK·

INTERNATIONAL
ADVERTISING ART

1929

PHÖNIX ILLUSTR·DRUCK & VERLAG·BERLIN SW 61 ✝

GIRDER

Für den sch

wε

anti

Schriftgießerei Ludwig

nen Katalog

elt-
qua

& Mayer Frankfurt a · M

Schräg laufende Zeilen fallen besonders im Anzeigensatz ins Auge. Das früher so mühsame Schrägsetzen bereitet Ihnen heute keine Schwierigkeiten mehr, wenn Sie unseren gesetzlich geschützten Schrägausschluß benutzen.

ABCDEGHI

KLMNOPU

Nr. 7278 84 Punkte
atz 34 Kilo 10 a 3 A
Kleiner Satz 20 Kilo 5 a 2 A

abcdefg

Geschichte des Barock

DIE KUNST

384 009   24 point large   6×A 11×a 7×1   1 fount about 6,5 lbs.

Printing Industry

384 005   12 point   15×A 26×a 11×1   1 fount about 4 lbs.

Tipos modernos de imprenta
NUEVOS ORLAS Y ADORNOS

384 010   30 point   5×A 10×a 5×1   1 fount about 8,25 lbs.

COCHE MOTOR

384 006   14 point   12×A 22×a 9×1   1 fount about 4,75 lbs.

The Master of the Press
MADE IN MANCHESTER

384 012   42 point   4×A 7×a 4×1   1 fount about 11,5 lbs.

Steamships

384 007   18 point   10×A 18×a 8×1   1 fount about 5,25 lbs.

New Railroad Stations
NEW TRANSMITTER

384 014   48 point   3×A 6×a 3×1   1 fount about 13,5 lbs.

POWERS

384 008   24 point small   8×A 14×a 8×1   1 fount about 6,5 lbs.

Nectares de frutas

Colours
for artists
and designers

withman

LONDON / NEW YORK / PARIS

384 015   60 point   3×A 5×a 3×1   1 fount about 20 lbs.

Bodega

384 016   72 point   3×A 5×a 3×1   1 fount about 27 lbs.

MARS

A B C D E F G H I J K L M N O P Q R S T U V W X Y Z
a b c d e f g h i j k l m n o p q r s t u v w x y z 1 2 3 4 5 6 7 8 9 0

Nr. 7265 6 Punkte Saß 3 Kilo 120 a 48 A

Groß und vielseitig ist das Erleben der menschlichen Seele und von gewaltigem Einfluß auf unsere ganze Wesensrichtung, auf den inneren Aufbau und die Entwicklung der Persönlichkeit. Kein Mensch wohl ist ganz empfindungslos den gewaltigen Eindrücken des

Nr. 7266 8 Punkte Saß 4,5 Kilo 94 a 38 A

Bei der Gestaltung der einzelnen Werbemittel wird sich der überlegene Kenner in jedem Falle auf ihre Bestimmung und die Geseße ihrer Wirkung besinnen. Das Plakat erfordert eine dem Inserat ganz

Nr. 7266 a 9 Punkte Saß 5 Kilo 70 a 28 A

Wenn wir von den Marionetten als Gleichnis des Lebens reden, ihr Spiel als dichterischen Ausdruck der Zeit erklären wollen, mag der Laie dem Thema skeptisch gegenüberstehen.

Nr. 7267 10 Punkte Saß 5 Kilo 70 a 28 A

Der Künstler, der harmonische Erzeugnisse irgendwelcher Art schafft, wobei wir nur den abgeglichenen Aufbau zu betrachten

Nr. 7268 12 Punkte Saß 6,5 Kilo 64 a 26 A

Die Ausstellung, die große Geländeflächen zu beiden Seiten der Seine bedeckte, war musterhaft ausgestattet

Nr. 7269 14 Punkte Saß 8 Kilo 56 a 22 A

Die Reklame als praktisches Werbemittel NEUE REKLAMESCHAU IN DRESDEN

Nr. 7270 16 Punkte Saß 9 Kilo 44 a 18 A

Die Musik im Dienst der Volksbildung KONZERTHAUS MARO HAMBURG

Nr. 7271 20 Punkte Saß 11 Kilo 40 a 16 A

Deutsche Kultur und Technik BUCHHANDLUNG DONHOF

Nr. 7272 24 Punkte Saß 12 Kilo 30 a 12 A

Sommer auf Westerland HOCHSEE-FISCHEREIEN

Nr. 7273 28 Punkte Saß 13 Kilo 20 a 8 A

Operetten-Schlager RICHARD TAUBER

Kunst und Literatur

Nr. 7274 36 Punkte
Saß 17 Kilo 18 a 6 A

Nr. 7275 48 Punkte
Saß 24 Kilo 14 a 6 A

Börsen-Zeitung

Kursbericht

Nr. 7276 60 Punkte
Saß 25 Kilo 10 a 4 A

Nr. 7277 72 Punkte
Saß 26 Kilo 10 a 3 A

Mercedes

Inventur

Nr. 7278 84 Punkte
Saß 34 Kilo 10 a 3 A

317

Schriftgießerei Ludwig & Mayer Frankfurt am Main

LYRA

PHOTO – 9174 · COLORIDA – *Lyra Orlow* ᛄ MADE IN GERMANY

PHOTO – 9174 · COLORIDA – *Lyra Orlow* ᛄ MADE IN GERMANY

PHOTO – 9174 · COLORIDA – *Lyra Orlow* ᛄ MADE IN GERMANY

PHOTO – 9174 · COLORIDA – *Lyra Orlow* ᛄ MADE IN GERMANY

PHOTO – 9174 · COLORIDA – *Lyra Orlow* ᛄ MADE IN GERMANY

PHOTO – 9174 · COLORIDA –

Lyra Orlow-COLORIDA - PHOTO - 9174 · MADE IN GERMANY

LYRA-ORLOW-Bleistiftfabrik NÜRNBERG

JOH. HA

KLISCHE

TEL. MORITZPLATZ 470   BER

ABA

Allgemeine Bau Ausstellun
Ausstellungshallen am Zoo
Vom 24. Januar–   8. Februar

RTLEIB
FABRIK
IN S 14 DRESDENERSTR. 34-35

nitt für Texteinsatz

Mörder
aller länder
vereinigt
Euch!

INTERNATIONAL ADVERTISING ART

Gebrauchsgraphik

1/1958   VERLAGSORT MÜNCHEN

GEBRAUCHSGRAPHIK

International Advertising Art

Herausgeber: Professor H. K. Frenzel, Editor ● „Gebrauchsgraphik" Druck und Verlag G. m. b. H.,
Berlin SW 61, Belle-Alliance-Platz 7-8 ● Alleinvertreter für die Vereinigten Staaten von Nordamerika
und Kanada: The Book Service Company, 15 East 40th Street, New York N. Y.-U. S. A. ● Sole
Representatives for the United States of America and Canada

Mai 1936.

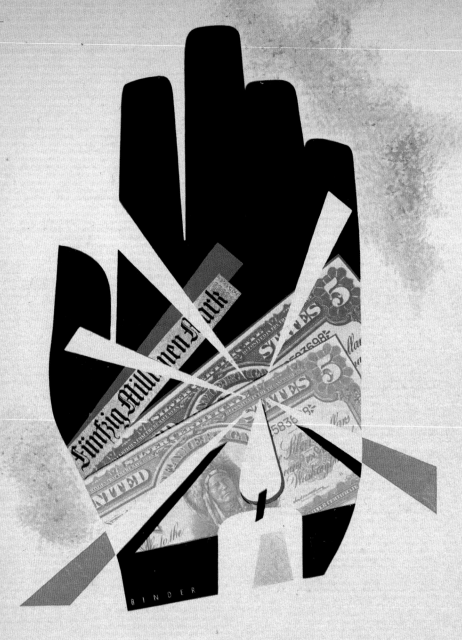

GRAPHIK

DIE ZEITSCHRIFT FÜR GEBRAUCHSGRAPHIK UND WERBUNG

10
1950

GRAPHIK

KONJUNKTUR

FORMGEBUNG

WERBUNG

DIE **MAPPE**

1

JANUAR 1948

326

DEUTSCHE MALERZEITSCHRIFT + DER SCHRIFTENMALER

DIE MAPPE

7

JULI 1959

DEUTSCHE MALERZEITSCHRIFT · DER SCHRIFTENMALER

WÄGEN UND WIRKEN

EIN DEUTSCHES LESE- UND LEBENSBUCH

VERLAG B. G. TEUBNER 1934

287. Point-of-purchase display for Othello pencils, *c.* 1932.

288–89. Specimen page of Egyptienne, *Schriften-Atlas*, Julius Hoffmann Verlag, Stuttgart, 1896.

290–91. Specimen page of Decorative, *Schriften-Atlas*, Julius Hoffmann Verlag, Stuttgart, 1896.

292–93. Specimen page of Italienne, *Schriften-Atlas*, Julius Hoffmann Verlag, Stuttgart, 1896.

294–301. Specimen sheets of Farbige Alphabete, designed by Otto Heim, *c.* 1932.

302. Magazine cover for *Gebrauchsgraphik*, July 1937.

303. Magazine cover for *Gebrauchsgraphik*, November 1937.

304–7. Specimen sheet of Farbige Alphabete, designed by Otto Heim, *c.* 1932.

308. Magazine cover for *Die Reklame*, August 1929.

309. Magazine cover for *Gebrauchsgraphik*, September 1929.

310–11. Cover for specimen catalog of Girder, Schriftgiesserei D. Stempel, date unknown.

312–15. Cover and interior pages for specimen catalog of Welt-Antiqua, Schriftgiesserei Ludwig & Mayer, 1932.

316. Specimen page of City Bold, Schriftgiesserei Ludwig & Mayer, 1930s.

317. Specimen page of Halbfette Welt-Antiqua, Schriftgiesserei Ludwig & Mayer, 1930s.

318–19. Point-of-purchase display card for Lyra pencils, *c.* 1935.

320–21. Letterhead and stickers, 1914–1925.

322. Magazine cover for *Gebrauchsgraphik*, designed by Eric Carle, January 1958.

323. Magazine cover for *Gebrauchsgraphik*, May 1936.

324. Magazine cover for *Graphik*, designed by René Binder, October 1949.

325. Magazine cover for *Graphik*, October 1950.

326. Magazine cover for *Die Mappe*, January 1960

327. Magazine cover for *Die Mappe*, July 1959.

328. Sample book jacket design from *Buch- und Katalogausstattungen, Inserate, Zeitungsköpfe*, from the series "Wie entwerfe ich Akzidenzen?", Willy Schumann, 1931.

SWISS

T HE POSTWAR SWISS GRAPHIC DESIGN THAT BEST CHARACTERIZED the International Typographic Style for which the country was known in the 1950s is firmly built upon a foundation of modern geometric sans serif types, like Akzidenz Grotesk, Univers, and Helvetica. Given this rigidly based typographic system, nineteenth- and early twentieth-century square slab serif faces were neither theoretically nor philosophically appropriate for the purist and reductivist aesthetics that the Neue Grafik designers sought to turn into a universal design formula. Nonetheless, some of the Modern movement's most influential practitioners used certain slab serif types in their work. Former Bauhaus member Max Bill's 1931 poster for "Negerkunst," an exhibition of prehistoric rock paintings of South Africa, and the jacket for Jan Tschichold's 1935 book *Typographische Gestaltung* (Typographic Design) were compositional tours-de-force in their applications of Georg Trump's angular slab, City Bold. For the cover of the first journal of the Swiss Typographic Organization, *Typographische Monatsblätter* (Typographic Monthly) in 1933, Walter Cyliax adapted older Antique slabs for a powerful constructivist composition (pp. 342–43).

The average Swiss commercial printers and typographers were no more immune to the lure of slab typefaces than their French, Italian, and German counterparts. When their choice was between serifs or sans, designers of everyday printed matter selected typefaces based on the power they had to turn a passerby's eye toward the message. Classic and custom slabs did that job well. Swiss type houses, wood type producers (pp. 332–37), and metal foundries made or imported tons of standard Antique and Clarendon styles. These, along with hand-drawn letters created for clients as varied as travel and liquor companies, meant that slab serifs were fairly widespread. They could be used in a variety of contexts, but they were most challenging and alluring when they were not totally in sync with the messages they were used to convey. The Steiner & Co. advert for printing "clichés" (p. 347) appears to contrast industrial-age letters with machine-age gears. This all goes to show that even in Switzerland, home of sans serifs, slabs put on a good face at every opportunity.

Série 6088

Classe C

36 Cic.

16 Cic.

RE

UNID

10 Cic.

BURGAR

12 Cic.

8 Cic.

Granada

Monde 253

6 Cic.

BORGOMARON

RH

Red

Montre 21

BORDEAUX

32 Cic.

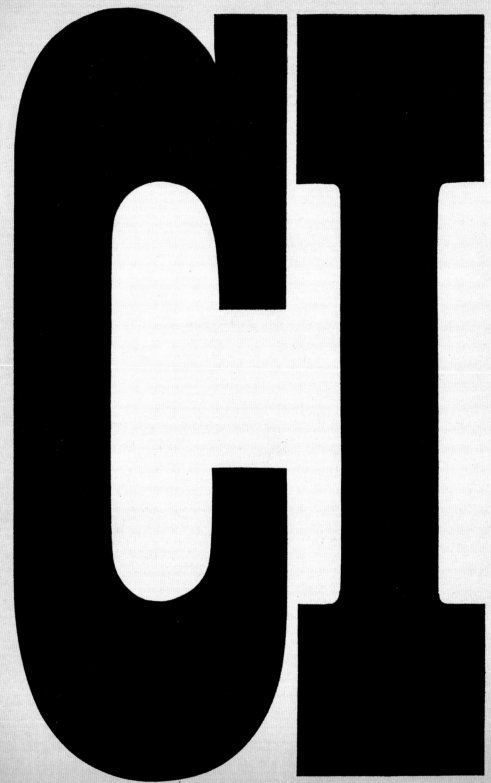

12 Cic.

BOIS

10 Cic.

Sonde

8 Cic.

GARNI

Série 6080   Classe F

20 Cic.

BE

16 Cic.

DIS

10 Cic.

Rand

12 Cic.

RED

10 Cic.

Mode

Cic.

GARDE

Cic.

Ostende 2

Cic.

IMPRIMEUR

357
American and Virginia Blend
KANSAS
AMERICAN CIGARETTES
Sw.Fr. 150
Sw.Fr. 150
KANSAS GOLD BLEND

Typographis M

che

onatsblätter

zur Förderung der Berufsbildung

Herausgegeben vom
Schweizerischen Typographenbund, Bern

Wäschepfl

Flo

pour l'entr

ege-Mittel

nis

tien du linge

CLICHES
STEINER & CO. BASEL

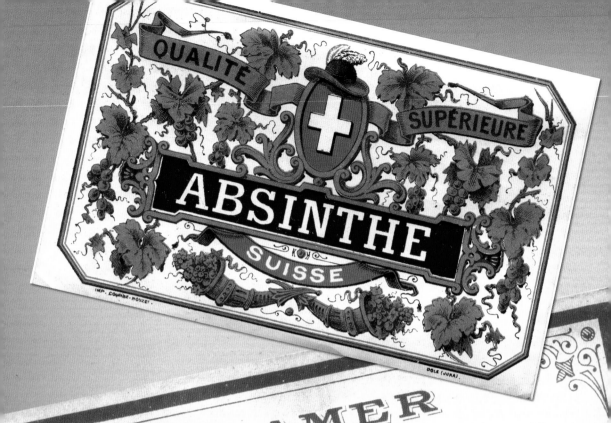

QUALITÉ

SUPÉRIEURE

ABSINTHE

SUISSE

IMP. COMBE-MOUZET.

DOLE (JURA).

VIN AMER

à base de VINS VIEUX et QUINQUINA

APÉRITIF

RÉCONFORTANT

CLOVIS REYMOND, VILLAMBLARD, (DORDOG

331. Liquor label, Vermouth Nugue-Richard & Cie., date unknown.

332–37. Wood type specimens, Roman Scherer, Lucerne, Switzerland, date unknown.

338–39. Travel brochure, "Été en Suisse," *c.* 1938.

340–41. Packages and counter display, *c.* 1938.

342–43. Detail of magazine cover for *Typographische Monatsblätter*, designed by Walter Cyliax, January 1933.

344–45. Package design for Floris, *Publicité et Arts Graphiques*, 1946–47.

346–47. Advertisements for Swiss printing firms, *c.* 1934.

348. Labels for Swiss liquor, dates unknown.

Annenberg, Maurice, *Type Foundries of America and their Catalogs*, New Castle, DE:
    Oak Knoll Press, 1994.

Clough, James and Chiara Scattolin, *Alphabets of Wood: Luigi Melchiori and the History
    of Italian Wood Type*, Treviso: Tipoteca Italiana, 2014.

DeNoon, Christopher, *Posters of the WPA 1935–1943*, Los Angeles, CA: The Wheatley Press, 1987.

Duvall, Edward J., *Modern Sign Painting*, Chicago, IL: Frederick J. Drake & Co., 1949.

Fili, Louise, *Grafica della Strada: The Signs of Italy*, New York, NY: Princeton
    Architectural Press, 2014.

Fili, Louise, *Graphique de la Rue: The Signs of Paris*, New York, NY: Princeton
    Architectural Press, 2015.

Harris, Elizabeth M., *The Fat and the Lean: American Wood Type in the 19th Century*,
    Washington, D.C.: National Museum of American History, 1983.

Heller, Steven and Seymour Chwast, *Graphic Style: From Victorian to New Century*,
    New York, NY: Harry N. Abrams, Inc., 2011.

Heller, Steven and Louise Fili, *Deco Type: Stylish Alphabets of the '20s and '30s*,
    San Francisco, CA: Chronicle Books, 1997.

Heller, Steven and Louise Fili, *Scripts: Elegant Lettering from Design's Golden Age*,
    London and New York, NY: Thames & Hudson, 2011.

Heller, Steven and Louise Fili, *Shadow Type: Classic Three-Dimensional Lettering*,
    London and New York, NY: Thames & Hudson, 2013.

Heller, Steven and Louise Fili, *Stencil Type*, London and New York, NY: Thames & Hudson, 2015.

Heller, Steven and Louise Fili, *Typology: Type Design from The Victorian Era to The Digital Age*,
    San Francisco, CA: Chronicle Books, 1999.

Heller, Steven and Louise Fili, *Vintage Type and Graphics: An Eclectic Collection of Typography,
    Ornament, Letterheads, and Trademarks from 1896 to 1936*, New York, NY: Allworth Press, 2011.

Hollis, Richard, *Graphic Design: A Concise History*, London and New York, NY:
    Thames & Hudson, 2001.

Hutchings, R.S., *A Manual of Decorated Typefaces*, London: Cory, Adams and Mackay, 1965.

Kelly, Rob Roy, *American Wood Type: 1828–1900: Notes on the Evolution of Decorated and Large
    Types*, New York, NY: Da Capo Press, 1969.

Lewis, John, *Printed Ephemera: The Changing Uses of Type and Letterforms in English and
    American Printing*, Woodbridge, Suffolk: The Antique Collectors Club, 1990.

Lewis, John, *The Twentieth Century Book: Its Illustration and Design*, London: Studio Vista, 1967.

McLean, Rauri, *Pictorial Alphabets*, London: Studio Vista, 1969.

McLean, Rauri, *Victorian Book Design and Colour Printing*, London: Faber & Faber, 1963.
    Enlarged edition, 1972.

Schriften Atlas, Berlin: H. Berthold, 1914.

Smith, Dan E., *Graphic Arts ABC, Volume 1: Square Serif*, Chicago, IL: A. Kroch & Son, 1945.

Specimen Book and Catalogue 1923, Jersey City, NJ: American Type Founders Company, 1923.

Spécimen général, Paris: Fonderie Deberny et Peignot, 1926.

Specimens of Printing Types, New York, NY: George Bruces's Son and Co., 1882.

Stephenson, Blake & Co., *Wood Letter*, Sheffield: Stephenson, Blake & Co., 1939.

Welo, Samuel, *Practical Lettering Modern & Foreign*, Chicago, IL: Frederick J. Drake & Co., 1930.

ACKNOWLEDGMENTS

MANY HEARTFELT THANKS TO LUCAS DIETRICH, our long-time editor at Thames & Hudson, for his continued support and invaluable guidance throughout this series. Gratitude to the following colleagues at T&H: Bethany Wright, Rosie Keane, and Johanna Neurath.

Thanks to the designers at Louise Fili Ltd, Nicholas Misani and Raphael Geroni. We could not have done this book without you.

We also thank Bill Moran at the Hamilton Wood Type and Printing Museum and Ross MacDonald at the Brightwork Press for their generosity in providing wood type proofs from their respective collections.

Although the majority of artifacts reproduced in this book are from our personal collections, they derive from various sources. We would most like to remember the late Irving Oaklander for his keen eye and wonderful stock of type and typography books.

Finally, we thank our son, Nicolas Heller. He is our inspiration for all things creative. S.H. *&* L.F.

FIN